# The Caribbean in the
# wider world, 1492–1992

# Geography of the World-Economy

*Series Editors:*

PETER TAYLOR   *University of Newcastle upon Tyne* (General Editor)
JOHN AGNEW   *Syracuse University*
CHRIS DIXON   *City of London Polytechnic*
DEREK GREGORY   *University of British Columbia*
ROGER LEE   *Queen Mary College, London*

A geography without knowledge of place is hardly a geography at all. And yet traditional regional geography, underpinned by discredited theories of environmental determinism, is in decline. This new series *Geography of the World Economy* will reintegrate regional geography with modern theory and practice – by treating regions as dynamic components of an unfolding world-economy.

*Geography of the World-Economy* will be a textbook series. Individual titles will approach regions from a radical political-economic perspective. Regions have been created by individuals working through institutions as different parts of the world have been incorporated in the world-economy. The new geographies in this series will examine the ever-changing dialectic between local interests and conflict and the wider mechanisms, economic and social, which shape the world system. They will attempt to capture a world of interlocking places, a mosaic of regions continually being made and remade.

The readership for this important new series will be wide. The radical new geographies it provides will prove essential reading for second-year or junior/senior students on courses in regional geography, and area and development studies. They will provide valuable case-studies to complement theory teaching.

# The United States in the world-economy
## A regional geography
John Agnew

*The United States in the World-Economy* is a major new textbook survey of the rise of the United States within the world-economy, and the causes of its relative decline. With the USA being the dominant state in the contemporary world-economy, it is vital to understand how it got where it is today, and how it is responding to the current global economic crisis. Professor Agnew emphasizes the divergent experiences of different regions within the USA and, in so doing, provides a significant 'new' regional geography, tracing the historical evolution of the USA within the world-economy, and assessing the contemporary impact of the world-economy upon and within it. No existing treatment covers the subject with equivalent breadth and theoretical acuity, and the guiding politico-economic framework provides a coherent radical perspective within which the author undertakes specific regional and historical analysis. *The United States in the World-Economy* will prove required reading for numerous courses in regional geography, area studies and the geography of the United States.

# South East Asia in the world-economy
Chris Dixon

South East Asia has for many centuries occupied a pivotal position in the wider Asian economy, linking China and the Far East with India and the Middle East and, since the early 1500s, the region has also played a major role in the world-economy. *South East Asia in the world-economy* is the first textbook survey of the area's interaction with these wider regional and international structures.

Professor Chris Dixon demonstrates how South East Asia's role has undergone frequent and profound change as a result of the successive emergence and dominance of mercantile, industrial and finance capital. He shows how the region has developed as a supplier of luxury products, such as spices; as a producer of bulk primary products; and how, since the mid-1960s, it has become a major recipient of investment and a favoured location for labour-intensive manufacturing operations, producing goods for European and American markets. The author examines how these phases in the evolution of the international economy have been reflected in the relations of production and in the spatial pattern of economic activity. He also discusses how the progressive integration of South East Asia in the world-economy has established the dominance of a small number of core areas and produced a pattern of uneven development throughout the region. In a concluding chapter, Chris Dixon explores the prospects for South East Asia in the 1990s in the light of the restructuring of the world-economy.

# The Caribbean in the wider world, 1492–1992

A regional geography

Bonham C. Richardson

*Professor of Geography, Virginia Polytechnic Institute and State University*

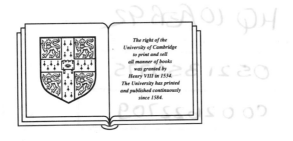

The right of the
University of Cambridge
to print and sell
all manner of books
was granted by
Henry VIII in 1534.
The University has printed
and published continuously
since 1584.

CAMBRIDGE UNIVERSITY PRESS

*Cambridge*
*New York Port Chester*
*Melbourne Sydney*

Published by the Press Syndicate of the University of Cambridge
The Pitt Building, Trumpington Street, Cambridge CB2 1RP
40 West 20th Street, New York, NY 10011-4211, USA
10 Stamford Road, Oakleigh, Melbourne 3166, Australia

© Cambridge University Press 1992

First published 1992

Printed in Great Britain at the University Press, Cambridge

*A Catalogue record for this book is available from the British Library*

*Library of Congress cataloguing in publication data*
Richardson, Bonham C., 1939–
The Caribbean in the wider world, 1492–1992: a regional geography
/ Bonham C. Richardson.
    p.    cm. – (Geography of the world-economy)
Includes bibliographical references and index.
ISBN 0-521-35186-3 (hardback). – ISBN 0-521-35977-5 (paperback)
1. Geopolitics – Caribbean Area.   2. Caribbean Area – History.
3. Caribbean Area – Economic conditions.   4. Caribbean Area –
Relations – Foreign countries.   5. Caribbean Area – Social
conditions.   I. Title.   II. series.
F2175.R49   1992
972.9 – dc20   91-9117 CIP

ISBN 0 521 35186 3 hardback
ISBN 0 521 35977 5 paperback

For Lucy Barbara Richardson

# Contents

# Maps

# Acknowledgements

My academic interest in the Caribbean region extends back nearly a quarter of a century to August, 1967, when I arrived in newly independent Guyana for field research for my Master's degree in geography at the University of Wisconsin at Madison. I returned to Guyana for a year in 1968–69 to research the Ph.D. degree at Wisconsin. Since then I have done field research in Trinidad (1971), Carriacou and Grenada (1973), St. Kitts–Nevis (1976), Barbados (1980 and 1981–82), and archival work dealing with the Commonwealth Caribbean in London (1986–87). I am very grateful to the American Philosophical Society, the Association of American Geographers, the National Geographic Society, The National Science Foundation, and the Society of the Sigma Xi for various grants that have helped fund my research. I also wish to thank Rutgers University and also Virginia Polytechnic Institute and State University for providing research support while I have held faculty positions at those universities.

During this more than twenty years I have augmented my primary evidence from field and archive by reading a good deal of the rich academic literature about the Caribbean. I have profited from the work of hundreds of writers, those living and dead and those from the Caribbean as well as those from elsewhere. Among the many, the writers who have been perhaps most influential for me are the geographer David Lowenthal and the anthropologist Sidney Mintz, two academics whose work I find myself perusing time and again for insight and inspiration.

Peter Taylor invited me to write this book and has encouraged me throughout, from the outline to the completed manuscript. He and Janet Momsen each provided helpful critiques of an earlier draft.

Jerome Handler directed me to pertinent photos at a crucial time. Patricia Mahoney created and drafted the maps in expert fashion. Vanessa Scott has again done heroic wordprocessing and has put up with more than she should from me. Thanks to The University of North Carolina Press for allowing me to use here parts of my essay about "Caribbean Migrations, 1838–1985," that appeared originally in Franklin W. Knight and Colin A. Palmer, *The Modern Caribbean* (1989).

Thanks also go to my wife Linda who, as a reference librarian, often has alerted me to pertinent publications and who has accompanied me in nearly all my field research. This book is dedicated to our younger daughter, without whom our lives would have been much duller and this book completed much sooner.

And thanks, perhaps most of all, to the many archivists, farmers, fishermen, government officials, market women, policemen, shopkeepers, teachers, and others in the Caribbean who have taken the time to share with me their observations and reflections about themselves, each other, and the rest of the world.

Bonham C. Richardson
Blacksburg, Virginia
October, 1990

# 1

## The creation of the Caribbean

The US Department of Commerce's 1989 guidebook, entitled *Caribbean Basin Initiative*, provides thumbnail sketches of commercial opportunities throughout the Caribbean for potential American investors. Haiti, the black republic occupying the western third of Hispaniola, is described therein as a particularly appealing locale: "Haiti's low wages, productive labor, strong private sector, and close proximity to the United States have been very attractive to offshore manufacturers, especially in electronics, apparel, toys, and sporting goods" (1988: 55).

The guidebook is not intended, of course, as an academic treatise, and it would be inappropriate to use it as a target. Yet its implications about Haiti (and very similar comments about other places in the circum-Caribbean region) are clear and important. Haiti's "low wages" and "productive labor" are portrayed as local cultural characteristics. And the happy combination of these indigenous Haitian characteristics with the "close proximity" of the world's largest national market suggests that it is not only economically rational but also helpful to all concerned that US manufacturers send component materials to these undemanding yet productive laborers for the fabrication of baseballs, playsuits, and TV parts for the North American market.

An historical-geographical assessment of Haiti provides a less exuberant but more instructive perspective. Aboriginal peoples called the entire island Quisqueya. It was "discovered," claimed for Spain, and renamed Hispaniola by Christopher Columbus in 1492. Nearly all the aboriginal peoples died or were killed off within the following three decades. In 1697 the western part of the island became French St. Domingue. France brought tens of thousands of

1

West African slaves to St. Domingue during the next century. Then the slaves rebelled against European control and established the independent republic of Haiti in 1804. Since then, the introduced overpopulation of African slaves and their descendants in Haiti have turned inward (partly because "civilized" countries refused to interact with Haiti for decades) to realize a near-hopeless case of overpopulation, ecological ruin, and poverty late in the twentieth century. Today's desperate pulsations of human migration from Haiti are attempts to escape this situation that in essence has been created by earlier European colonial strategies. One of the few local sources of wages is sweatshop fabrication of "electronics, apparel, toys, and sporting goods" in makeshift factories whose products are destined for the United States. And Haitians know full well that clamoring for higher wages or striking for better working conditions probably would send the Americans packing in search of more hospitable environs.

Haiti's particular characteristics are, of course, unique, but the general contours of her cumulative experiences of externally imposed underdevelopment have parallels elsewhere in the Caribbean and, in a broader sense, throughout the Third World. During the past quarter-century social scientists have begun to develop theoretical perspectives in attempting to explain these perceived regularities. Among the most prominent has been the sociologist-political scientist Immanuel Wallerstein's historical assessment of the origins and development of a European-centered world-economy, driven by capitalism and manifested by an international division of labor. Wallerstein asserts that since the late 1400s the genesis of the capitalist world-economy has led to the formation of "core areas" (which now include North America) and the outlying "periphery" (including Haiti and the rest of the Third World). Core and periphery are intimately linked by unequal economic exchanges, and low wages in the periphery (as opposed to high wages in the core) constitute a fundamental and enduring characteristic of the world-economy (1974).

It is not surprising that Wallerstein's (here very oversimplified) thesis, and its many progeny, has created intense debate and considerable excitement throughout the social sciences. If nothing else, the now ubiquitous world-economy outpouring has laid to rest the "strange peoples and places" way of viewing various other parts of the world. Scholars now look beyond the village walls or local gardens of Third World peoples to seek explanations as to how external events affect and are affected by village inhabitants. And these villagers, it is important to note, are no longer considered passive recipients of external stimuli but agents who are part of the global economy's overall trajectory. Among others, academic geographers have been enthused by the world-economy perspective (e.g., Taylor 1986, 1988). Geographers, whose long-held predilections with places and regions were overshadowed by spatially or geometrically oriented colleagues in the 1960s and 1970s, are beginning to re-discover places and regions through this world-economy perspective and therefore to reassess particular areas of the world in an entirely different way (e.g., Agnew 1987). Further, there are unmistakable diagnostic indications that the world-economy perspective has taken hold in academic geography, notably with the arrival of textbooks dealing with the idea (e.g., Knox and Agnew 1989) and an incipient countercurrent of literature among geographers objecting to the world-system framework as an analytical device (e.g., Harvey 1987).

In this book I attempt to view the Caribbean in a world-economy perspective. The term "perspective" is crucial because there is no attempt here to adhere so closely to a world-economy model so as to provide, for example, a theoretical comparison of Caribbean economic trends or cycles with waves and curves that others have perceived in the world-economy as a whole over the centuries. Rather, I attempt to show the ways in which external control of the Caribbean region has influenced landscapes, ecological problems, settlement forms, demographic characteristics, migration patterns, livelihood strategies, and other variables within the Caribbean – in other words, issues that have been traditional themes in academic geography. Although the emphasis here is geographical, I also emphasize the history of the region and its component parts throughout. The historical emphasis is not provided as obligatory and passive "background" material but as a concrete historical experience that informs the present and which in many ways continues to provide meaning and significance for the people who inhabit the Caribbean region today, very late in the twentieth century.

It is probably accurate to suggest that world-economy thinking in general has not been viewed as nearly so novel by Caribbeanists as it has by, say, academic specialists dealing with Africa or Asia. This is because of the obvious and thoroughgoing domination of the Caribbean region for centuries by external powerholders who have transformed landscapes and local populations to meet outside market needs. Long before world-economy or "dependency"-type concepts were common currency among social scientists, a considerable literature – academic and popular – recognized the outward-focused character of the region. Caribbeanist academics now routinely use the world-economy framework to analyze contemporary economic data from the region (e.g., Ramsaran 1989). And Caribbean peoples have lamented for decades, as they do today, that local success often is achieved by those emigrating to North America or Europe and then returning, rather than by those who stay behind.

Much of the reason for the Caribbean's external focus is its longevity as an appendage of the world-economy. Columbus's voyage at the end of the fifteenth century brought the Caribbean into Europe's orbit as its first overseas colonial outpost. But for the next century and a half the Caribbean served essentially as a Spanish transit zone between Spain and its populous and mineral-rich colonial empire in Central and South America. The Caribbean region itself was not irrevocably harnessed to the European economy as a whole until the mid-seventeenth century, a harnessing that took place during the "long contraction" of the world-economy between 1600 and 1750 and because the Caribbean could produce tropical staples, mainly sugar cane, that could not be cultivated in Europe (Galloway 1989: 78–79; Hobsbawm 1967: 53–56; Wallerstein 1980: 166–67). Since that time the Caribbean has been closely linked with the European and, later, North American-centered world-economy, experiencing the innumerable and often damaging effects of international capitalism's advances and retreats, swings in commodity prices, and resultant booms and busts.

Over the centuries the Caribbean region has therefore been a geographical receptacle for a diverse flow of material items and cultural stimuli from outside the region – crops, weeds, animals, peoples, technology, food items, ideas, and much more. These variables have been absorbed, modified, and transformed in characteristic ways so that one may speak of regional "Caribbean" examples of, for example, ethnic identity, crop combinations, or settlement patterns. Yet it is crucial to note that regionality as expressed by regional characteristics in the Caribbean is an abstraction and perhaps more so than in other broadly delineated world regions. Within the Caribbean "regional" matrix, imported and local geographical variables have combined in a great many ways in different places so that the Caribbean is in reality a cultural mosaic of subtle complexity and incredible variety (Lowenthal 1960b); regularities identified in one Caribbean locale – to the chagrin of those who would seek broad regional generalizations – are often absent in the next. And those seeking explanations for this complex regional variety would probably find their most satisfying answers in exploring interactions among the region's lengthy colonial history, the variety of competing powers that have influenced the region, and the physical fragmentation of the Caribbean area itself.

The Caribbean's long association with Europe and its interaction with other world regions suggest that cross-regional comparisons might be fruitful. Indeed, one finds European housetypes, Asian religious edifices, and Africanisms of all sorts throughout the region. But – contrary to tourist brochures – Barbados is not a "Little England," Fort-de-France (in Martinique) is not a miniature Paris, and St. Maarten is vastly different from anything found in the Netherlands. Nor is the Caribbean, despite its geographical proximity, simply a neglected appendage of "Latin America" because, in world-economy parlance, the Latin American mainland was not "integrated into European intent at the same time, at the same rates, in the same ways, or with the same results as the Caribbean islands and their nearest mainland surroundings" (Mintz 1977: 254).

Although this book emphasizes for the most part the effects of external stimuli on the Caribbean region over the past half millennium, the Caribbean has reciprocated by affecting the internal character of the dominant metropolitan countries themselves. During the colonial plantation era, labor regimens on Caribbean sugar-cane plantations imposed a regimented, "industrial" routine on enslaved work forces well before the Industrial Revolution, as a kind of preview of what Western European workers later would experience. And the sugar produced by Afro-Caribbean slaves sweetened the coffee and tea that were important dietary supplements of Europe's burgeoning proletariat during the nineteenth century (Mintz 1985).

More immediately, in the late twentieth century emigration from the Caribbean region has brought hundreds of thousands of Caribbean peoples to the doorsteps of the colonial and neocolonial nations that have historically created the conditions encouraging these migrations in the first place. These newcomers have altered the fabric of their "host" countries by introducing their own foods, clothing, celebrations, work ethics, and zeal for education. Much more important, the somewhat sudden presence of large numbers of Caribbean peoples in North Atlantic metropoles provides the basis for serious self-reflection. For Americans who ostensibly pride themselves on their historic assimilation of peoples from throughout the world, the immigration of hundreds of thousands of Afro-Caribbean peoples into the United States represents at once a test of

ability truly to welcome and assimilate as well as an extension of a
more numerous Afro-American presence that is centuries old. For
Europeans, Caribbean migration represents a geographic reversal of
an issue as old as empire itself: the associations between rulers and
ruled, abstractions formerly confined to the far corners of the globe,
are now, owing largely to the presence of Caribbean peoples, local
European issues of immediate significance (Richardson 1989).

It should come as little surprise that, given outsiders' domination
of the Caribbean region for centuries, external perceptions of the
region also have contributed to the confusion of just what is meant
by "Caribbean" (Map 1). No one would disagree that the region's
core consists of the arcuate archipelago of islands that stretches from
the Yucatán and Florida peninsulas southeast to Venezuela, with the
Greater Antilles (Cuba, Hispaniola, Puerto Rico, Jamaica) in the
north and the smaller islands, or Lesser Antilles, generally to the
south and east. This book also includes the Guianas, on the northeastern coast of South America, as "Caribbean" because of their
close historical and cultural ties to the region (Lewis 1983: 2–3) and
also the Central American rim, notably Belize, for the same reasons
(Augelli 1962). As in the delineation of any large cultural region,
these vague boundaries are fuzzy, permeable, and somewhat arbitrary.
Beyond brief discussion of the noticeable Afro-Caribbean ethnic
presence along the eastern lowlands of Nicaragua, Costa Rica, and
Panama, for example, I choose not to include these countries within
the Caribbean realm, as they are more conventionally "Central
America." Yet these Central American nations, as an instance of how
far the Caribbean region can extend according to some observers, are
included in the United States' Caribbean Basin Initiative program.
As a further example, one could make a strong case for including
southern Florida as part of the Caribbean region on the basis of the
immense cultural impact of the relatively recent arrival of hundreds of
thousands of Caribbean peoples in the Miami area.

The term "West Indies" is roughly synonymous with "Antilles"
and refers to the islands themselves, including the Greater and Lesser
Antilles. "West Indies" is thus more restrictive than "Caribbean,"
and there is disagreement as to whether the Bahamas and the Dutch
islands west of Trinidad are part of the West Indies. Residents of
any of the islands are usually referred to as "West Indians" by most
local English-speakers. Further confusion may be provided by noting
that Belizeans and Guyanese are much more likely to be dubbed
"West Indian" than are Spanish-speaking residents of Cuba, the
Dominican Republic, and Puerto Rico.

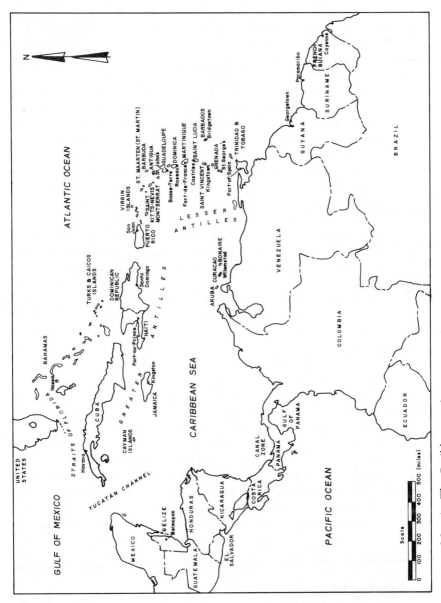

Map 1  The Caribbean region

Lightly populated when compared to most other world regions, the Caribbean has roughly 33 million human residents (Table 1). Ethnically, the region is heavily black with perhaps two-thirds of its people of black African descent or with black African admixture. The preponderance of black peoples in most Caribbean locales is, of course, a legacy of the African slave trade of colonial times. And large "minority" segments of the Caribbean kaleidoscope of peoples – such as the Portuguese and Asian Indians of the southern Caribbean and many of Cuba's Spaniards – came as plantation workers after slavery. As they are elsewhere, skin color, social status, and the potential for economic prosperity are intimately bound up with one another in Caribbean locales. Skin color and "class" are, however, not always easily correlated with one another, and the many variations pertaining to ethnic identity and social status within the Caribbean region have spawned a massive academic literature (e.g., Hoetink 1967; Lowenthal 1972).

ϒ The Caribbean's cultural complexity is, however, perhaps more evident in its spectrum of spoken languages than in its array of skin colors. Spanish is the lingua franca for roughly 60 per cent of Caribbean peoples, the vast majority of them in Cuba, the Dominican Republic, and Puerto Rico. French is the language of France's three overseas departments, a French-based creole is the spoken and written language of Haiti, and most Dominicans (of Dominica) and St. Lucians speak a French-based language as well. Residents of the other former British colonies are English-speakers, although a few very old Trinidadians and Guyanese speak only Hindi. Dutch is not as common in the Netherlands Antilles; many residents of those islands speak a mixture of English, Dutch, and Spanish known as Papiamento. Aboriginal languages survive in the back country of the Guianas alongside the language used by the Suriname "Bush Negroes" who are descendants of escaped slaves and who converse with one another using many words of African derivation. Nuances from one island to the next in spoken languages ("Cuban" Spanish, "Jamaican" English) are noticeable to residents of the Caribbean but difficult for many visitors to detect. Many peoples of the Caribbean region are bilingual, especially in that the language they speak at home is very different from that same language they speak to outsiders. American television, some of it dubbed into local languages but much of it with its original American English, creates a considerable linguistic impact throughout the Caribbean late in the twentieth century.

I have written this book over a span of somewhat more than two years, a task interspersed with regular teaching duties and along with the preparation of other articles, papers, and reviews. A certain anxiety accompanies the preparation of any comparable book-length study because the author must be aware of material published during the writing that requires a rewriting or even rethinking of the work in progress. I have attempted to assimilate ideas from much of the very recent material dealing with the Caribbean appearing during my writing, although I am certain that some of it has escaped me. Similarly, and of even greater concern, is that I have written a book that is ultimately intended to portray the current character of the Caribbean region. Any such attempt is automatically unsuccessful to a degree because the lag time between the completion of a manuscript and the book's eventual appearance means that the latter will inevitably be something of an instant anachronism.

These points are gratuitous but nonetheless important for a book dealing with an entire region of the world. Their importance is underlined with a brief, selected list of particular events that have occurred and are occurring in the Caribbean – and very close to it – during the writing of this book: two major hurricanes have ravaged different parts of the region; Haiti's political regime has been overthrown twice; Michael Manley has defeated Edward Seaga to become Jamaica's Prime Minister, a reversal of the results of Jamaica's 1980 election; Puerto Rican momentum grows for the island to become the 51st US state; the American crackdown on the drug traffic throughout the region has had its most dramatic expression in the invasion of Panama in December, 1989; and the astounding political transformations in Eastern Europe combined with severe local economic decline lead many to wonder how much longer Fidel Castro can maintain control in Cuba.

Paralleling the headlines, a series of underappreciated yet geographically more important events also have taken place in the Caribbean during the writing of this book; they have less to do with recent sociopolitical changes and more to do with the region's continuity with its past: tens of thousands of babies have been born; thousands of young men and women have decided to emigrate; families have coped with rising prices of imported lumber and foodstuffs; seasonal tourists from Europe and North America have descended and then departed; small-scale farmers have calculated crop combinations in light of their subsistence needs, fluctuating external commodity prices, and local ecological characteristics. These everyday events are, of course, not unrelated to the more newsworthy items mentioned in the preceding paragraph. Nor can

Table 1. *Caribbean country data*

| | Political status | Area in square miles | Population (000s) | Capital city | Life expectancy | Per capita income (US) | Remarks |
|---|---|---|---|---|---|---|---|
| Anguilla | British Colony | 35 | 7 | The Valley (1,000) | 71 | — | tourism, shrimp |
| Antigua and Barbuda | Independent (1981) | 171 | 83 | St. John's (30,000) | 70 | 2,380 | tourism, oil refining |
| Aruba | Non-metropolitan territory of the Netherlands | 75 | 60 | Oranjestad (19,800) | 74 | 2,350 | tourism, banking |
| Bahamas | Independent (1973) | 5,382 | 245 | Nassau (110,000) | 69 | 7,190 | tourism, banking |
| Barbados | Independent (1966) | 166 | 254 | Bridgetown (102,000) | 73 | 5,140 | tourism, sugar cane |
| Belize | Independent (1981) | 8,867 | 178 | Belmopan (3,500) | 69 | 1,170 | citrus, sugar cane |
| British Virgin Islands | British Colony | 59 | 12 | Road Town, Tortola (2,500) | 71 | 8,170 | tourism, government employment |
| Cayman Islands | British Colony | 102 | 23 | George Town (11,500) | 62 | 12,900 | tourism, banking |
| Cuba | Independent (1902) | 42,804 | 10,421 | Havana (2,037,000) | 75 | 2,690 | sugar cane, nickel |
| Dominica | Independent (1978) | 290 | 79 | Roseau (8,300) | 76 | 1,210 | limes, bananas |
| Dominican Republic | Independent (1844) | 18,704 | 6,850 | Santo Domingo (1,410,000) | 63 | 710 | sugar cane, coffee |
| French Guiana | Overseas Department of France | 33,399 | 92 | Cayenne (37,000) | 67 | 2,340 | shrimp, French space station |
| Grenada | Independent (1974) | 133 | 106 | St. George's (7,500) | 67 | 1,240 | bananas, spices |
| Guadeloupe | Overseas Department of France | 687 | 340 | Basse-Terre (13,000) | 73 | 3,300 | sugar cane, tourism |
| Guyana | Independent (1966) | 83,000 | 757 | Georgetown (200,000) | 69 | 500 | sugar cane, bauxite |

| | | | | | | | |
|---|---|---|---|---|---|---|---|
| Haiti | Independent (1804) | 10,597 | 5,451 | Port-au-Prince (473,000) | 53 | 330 | coffee, light manufacturing |
| Jamaica | Independent (1962) | 4,244 | 2,407 | Kingston (525,000) | 70 | 880 | bauxite, sugar cane |
| Martinique | Overseas Department of France | 421 | 336 | Fort-de-France (97,000) | 74 | 4,280 | sugar cane, bananas |
| Montserrat | British Colony | 40 | 12 | Plymouth (12,000) | 71 | 2,350 | tomatoes, tourism |
| Netherlands Antilles | Non-metropolitan territory of the Netherlands | 308 | 177 | Willemstad, Curacao (125,000) | 73 | 6,020 | oil refining, banking |
| Puerto Rico | United States Commonwealth | 3,515 | 3,301 | San Juan (429,000) | 75 | 5,190 | tourism, manufacturing |
| St. Kitts and Nevis | Independent (1983) | 104 | 43 | Basseterre (18,500) | 65 | 1,700 | sugar cane, tourism |
| St. Lucia | Independent (1979) | 238 | 145 | Castries (53,000) | 71 | 1,320 | bananas, tourism |
| St. Vincent and the Grenadines | Independent (1979) | 150 | 113 | Kingstown (19,000) | 69 | 960 | bananas, food crops |
| Suriname | Independent (1975) | 63,251 | 425 | Paramaribo (68,000) | 68 | 2,510 | shrimp, rice |
| Trinidad and Tobago | Independent (1962) | 1,978 | 1,258 | Port-of-Spain (57,000) | 69 | 5,120 | petroleum, sugar cane |
| Turks and Caicos Islands | British Colony | 193 | 11 | Cockburn Town (3,500) | 70 | 4,490 | tourism |
| US Virgin Islands | United States Territory | 136 | 107 | Charlotte Amalie, St. Thomas (12,000) | 69 | 10,050 | tourism, shipping |

Data from *Encyclopaedia Britannica 1989 Book of the Year*.

these prosaic occurrences really be classified as indices of "continuity" while the headlines indicate "change" because several of the latter events – including political upheaval, environmental catastrophes, and gunboat diplomacy – are recurrences of long standing in the Caribbean's region.

Together, the continuity exhibited by both grassroots and broadly political events in the region would seem to justify the long view of the Caribbean which is presented in this book. The book's organization is similar to other regional geographies. A chapter that deals with the Caribbean's physical environment is followed by two historical chapters. The remainder of the book discusses the region's present as it has been informed by the past. A slight overlap among the chapters is intended, although repetition is unintended. The chapter dealing with resistance is, to my knowledge, novel among regional geographies of the Caribbean. Yet the resistance theme is vital lest the world-economy perspective suggest a portrayal of Caribbean peoples as faceless automatons energized only by metropolitan stimuli, rather than the adaptive and creative peoples that they truly are.

# 2

## A colonized environment

Our landscape is its own monument: its meaning can only be traced on the underside. It is all history.

Edouard Glissant, *Caribbean discourse* (1989)


A few hours inland from the Haitian capital city of Port-au-Prince, a black rural cultivator named Nesmère controls a three-acre plot of land. Like other rural Haitians, and like thousands of other small-scale farmers throughout the islands of the Caribbean Sea, Nesmère produces a combination of cash and subsistence crops. The beans, yams, corn, breadfruit, guava, and citrus are for Nesmère, his wife, and their three children. The coffee that he cultivates in the shade of the breadfruit and citrus trees finds its way into an international commodity flow through a complex local marketing system (Grove 1981; Girault 1985).

Nesmère, his family, and his neighbors are, of course, vitally dependent on their immediate physical environment for life and livelihood. So when Hurricane Allen ravaged the southwestern peninsula of Haiti in 1980, rural Haitians went even hungrier than usual, and some survived only because of food relief supplies from the United States. But weather hazards, including hurricanes and droughts, are only ephemeral events. The fundamental ecological problem in Haiti is its soil erosion. Too many people for too little arable land has pushed Haitian subsistence cultivation onto ever-steeper hillsides, creating an all-too-familiar vicious circle in which the land's capacity to produce is continuously undermined by the necessity to cultivate marginal slopes. During periods of drought – whose effects are intensified by a lack of protective vegetation – parts of southwestern Haiti appear barren and bleak; the topsoil has long since been washed away, leaving sun-bleached bedrock where subtropical forests once stood (Grove 1981: 246).

Haiti is perhaps the poorest country in the Western Hemisphere. Extending observations from Haiti to the Caribbean region as a

13

whole must therefore be done with caution. Yet the historical background of Haiti's contemporary ecological dilemma is typically Caribbean. The aboriginal peoples of the region farmed and fished and modified the insular landscapes relatively little. Then centuries of European exploitation (mainly by France in Haiti's case) were marked by massive environmental change. West African slaves, laboring under European supervision, stripped the natural vegetation from Caribbean islands and planted tropical staple crops – mainly sugar cane – in its place. Heterogeneous insular ecosystems were thus transformed into monocrop economies. Local environmental corrective actions were negligible because the decisions affecting the Caribbean environment were made by external market forces and implemented by transient European planters. Massive local soil erosion followed. Introduced overpopulation exerted additional stresses on insular environments during the slavery era and beyond. The Caribbean's environmental degradation and associated land–people imbalance of the late twentieth century are thus direct legacies of the region's half millennium of outside control.

Political independence for most places in the Caribbean region in the late twentieth century has not ended external influences on Caribbean environments. The region's agricultural and mineral exports are still almost entirely dependent on the metropolitan markets of Europe and North America. And even outsiders' images of Caribbean landscapes – the sun, sand, and sea stereotypes portrayed in European and North American tourist brochures – are crucial to the economic health of the Caribbean region. Throughout recorded history, and including the present, the economic benefits from Caribbean environments, whether commodity production or scenery, have therefore accrued "not to inhabitants but to outsiders" (Lowenthal 1987a: x).

Even the sharply delineated political boundaries that insularity – the Caribbean's geographic hallmark – usually provides have been subdivided by external rivalries for the region. The political boundary on Hispaniola between (formerly French) Haiti on the west and the (formerly Spanish) Dominican Republic is only the most obvious example. The Dutch and French continue to share tiny St. Martin, the French and British jointly occupied St. Kitts until 1713, and changing geopolitical fortunes have complicated most individual island cultures with stratified or "layered" historical pasts that continue to influence the present.

## The physical background

The scenic allure of the Caribbean region is enhanced because of the mountainous character of many of the islands. Even some of the smallest islands have peaks soaring to 4000 ft (over 1200 m.) which are obscured most afternoons by mist and clouds owing to the heat and humidity of the subtropical atmosphere. Yet there are many low-lying islands as well, and probably the most appropriate generalization concerning the landform geography of the Caribbean realm is that it is characterized by surprising diversity, surprising because of the monotonous sun-and-sand television advertisements beamed into mid-latitude households portraying the Caribbean region simply as a beach.

The complex mountain ranges of the Greater Antilles are structurally associated with the geological sub-region called "Old Antillia", whose formation dates back approximately 100 million years to the Cretaceous period in geologic history (Watts 1987: 5–11). After their volcanic formation, the "proto-Antilles" were then below the ocean surface and subjected to submarine depositions of both sandstones and limestones and also to intrusion from below by enormous granitic batholiths. Today the highest mountains of the Greater Antilles are underlain by the batholiths as well as complexes of altered sandstones: the Sierra Maestra of extreme southeastern Cuba, 6470 ft. (1972 m.) at the highest point; the Blue Mountains of central Jamaica which rise to 7405 ft. (2257 m.); and the highest mountain in the Caribbean, Pico Duarte at 10,417 ft. (3175 m.) in the Dominican Republic.

The historical importance of the mountains of the Greater Antilles is far greater than that of simply providing scenic but passive backdrops for the unfolding of Caribbean history. The inaccessible limestone hills and mountains of central Jamaica were highland sanctuaries that protected the runaway slaves ("Maroons") from coastal planters and provided the physiographic underpinning for a remarkably persistent independence from plantation oppression well before freedom came to the slaves on coastal estates. During the 1950s, the Sierra Maestra range furnished a refuge for Fidel Castro's hit-and-run raids against numerically superior Cuban armed forces until his followers toppled the Batista regime in 1959.

The Greater Antilles also have the most extensive areas of plains or flatlands in the Caribbean region. The dark-red clays of the plains of central and western Cuba are perhaps the most fertile soils of the Antilles, having supported crops of sugar cane for centuries. And the

Cibao lowland depression in the western section of the Dominican Republic, together with the plain in the southeastern quadrant of the same country, are major agricultural zones.

The arc of the Lesser Antilles, curving south from St. Martin in the north to Grenada in the south, lies along the intersection of two of the earth's crustal plates. The resultant vulcanism has produced a chain of mountainous islands whose scenic beauty awes visitors and residents alike. Volcanic peaks soar almost from the water's edge in several of the islands, the highest mountain in the chain being Morne Diablotine on Dominica, at 5155ft. (1568m.) in elevation. Erosion from the peaks has produced aprons of reasonably fertile soil around the mountains of each island, although the volcanic Lesser Antilles are by no means uniform physiographically. St. Kitts, one of the northernmost of the Lesser Antilles, for example, has excellent ash soils that have supported sugar-cane crops continuously for three and one-half centuries. Nevis, only two miles south across a narrow channel from St. Kitts, has a higher clay content in its soil, subsequently higher erosion rates, and a generally steeper and stonier landscape (Lang and Carroll 1966).

The rugged scenery of these small volcanic islands is matched by their frequent seismic tremors and shocks, well-known to anyone who lives or visits there. The archival colonial records from each of these places abound with earthquake reports, giving full descriptions of flattened churches and ruined colonial buildings throughout the region's colonial history. And rural peoples in each of the islands have a complementary stock of personal anecdotes about earthquakes and landslides. On June 7, 1692, a massive quake sent Port Royal, Jamaica – a notorious haunt of buccaneers, prostitutes, and predatory colonial administrators – sliding into the sea (Dunn 1972: 186–87). The most renowned volcanic eruption in recent Caribbean history was that of Mt. Pelée in Martinique in May, 1902: A combination of lava and superheated gases from the volcano killed more than 30,000 persons in the capital city of St. Pierre and also destroyed vessels moored in the harbor nearby. Ominous rumblings combined with comparatively minor eruptions in the late twentieth century, notably in St. Vincent and Guadeloupe, serve as ongoing reminders that a similar catastrophe could occur again at any time (Sigurdsson 1982).

East of the volcanic arc of the Lesser Antilles lies a parallel line of low-lying, coralline islands, including Barbuda, Antigua, the eastern half of Guadeloupe, and also Barbados. The absence of numerous well-defined streams on these islands eliminates muddy fluvial discharges, thereby enhancing the picturesque beaches with obvious implications for these islands' important tourist industries.

Trinidad and Tobago, the southernmost of the Lesser Antilles, are composed primarily of sedimentary rocks and geologically are more closely associated with South America than with the volcanic island chain stretching north. Roughly six hundred miles west of Trinidad are three of the Netherlands Antilles, Aruba, Bonaire, and Curaçao – the so-called ABC islands – off the coast of Venezuela. The infertile soil of these three are derived from an ancient crystalline base, although these islands' soil capabilities are of much less economic importance than are their strategic locations.

The general physical similarities between the eastern rim of Central America and the Guianese coastal plain help to explain in an indirect way why these areas usually are considered culturally Caribbean. Mudflats, mangroves, and a relative dearth of pre-European peoples rendered these swampy coastal zones unattractive to the thrust of earliest Spanish colonization on the mainland, which was focused on both gold and the domination of sizable aboriginal populations. So the control of these coastal backwaters on the "Latin" mainland was left for colonial latecomers – the British, Dutch, and French – who helped to create a heterogeneous ethnic mix there with imported labor forces.

Although these two coastal fringes of the Caribbean region both are low-lying littorals, they actually are very different geomorphologically. Sluggish rivers flowing eastward from the Central American mountains deposit their muddy loads in the coastal zone of Belize which is fringed with innumerable coral and limestone reefs. Westward, beyond a series of ancient pine ridges, the hills of central Belize are similar to neighboring areas of Guatemala (West and Augelli 1976: 415). The fine-particle clays of the Guiana coastal plain, in contrast, are carried all the way from the mouth of the Amazon River (Richardson 1987). An equatorial current periodically erodes and at other times enhances the coastal mudflats of French Guiana, Suriname, and Guyana, the heaviest volumes of sediment usually deposited along the foreshore of the latter country. Guyanese of the late twentieth century inhabit a semi-aquatic settlement matrix originally laid out by Dutch and English planters; and present-day Guyanese, as one result, are known as "mudlanders" to their insular West Indian neighbors.

A brief dry spell in late summer on the Guyanese coastal plain is evident most years because of prevailing southeasterly tradewinds that parallel the coastline. Georgetown, Guyana's capital city, lies at

7 degrees north latitude, and Paramaribo in Suriname and Cayenne in French Guiana are slightly south of that, so tradewinds originating in the southern hemisphere affect all the Guianas during late summer when the intertropical convergence zone (marking the meeting or convergence of air masses from the northern and southern hemispheres) shifts to the north.

But the coastal zone of the Guianas is the only part of the Caribbean realm not under the year-round influence of the northeasterly tradewinds that blow westward from the subtropical high pressure system (or "Azores high") of the north central Atlantic. Speaking very generally, most of the rain that falls on the Caribbean islands and adjacent areas of eastern Central America comes in mid to late summer as the Atlantic high pressure cell is at its farthest point north. Easterly winds during this period must thus traverse hundreds of miles of warm, subtropical, ocean surface before reaching the islands, by which time the moving air masses are laden with atmospheric moisture.

When the Atlantic high pressure cell begins to shift south, in late June and July, the movement helps to generate complex atmospheric perturbations off West Africa resulting in westward-moving low-pressure storms which, on occasion, develop into hurricanes. These late-summer storms are best known, of course, for their occasional furious visitations to particular Caribbean islands, leaving death, damage, and devastation in their wakes. Yet seasonal, hurricane-related rains have left a more prosaic meteorological signature in the region in the form of the well-known "hurricane curve," the lopsided histogram that represents the heavily skewed rainfall distribution from July to October throughout the region, from Barbados to Belize, to the Bahamas.

Hurricanes have relentlessly punctuated Caribbean autumns with seasonal predictability but with widely varying paths or trajectories. In general, they enter the region from the east through the Lesser Antilles and depart toward the north. Beyond that vague generalization, their routes are nearly impossible to predict. But modern technology now tracks these late summer storms to the benefit of all residents of the Caribbean region. During the earliest European occupation of the Lesser Antilles, planters received only minimal warning of approaching hurricanes from the native Caribs (who they then accused of consorting with the devil when the hurricanes did come). Forecasting hurricanes improved only slightly until the mid-twentieth century. Today, late in the twentieth century, stationary satellites monitor the paths of hurricanes across the Atlantic from their spawning grounds off West Africa, providing advance storm warnings that are broadcast on Caribbean radio and television stations.

But sophisticated forecasting procedures are neither infallible nor can they really provide protection against the massive autumn storms. In September, 1988, Hurricane Gilbert, described as "the mightiest storm to hit the Western Hemisphere in this century" (Browne 1988) did not assume hurricane force until it was west of the Lesser Antilles; Gilbert then raked the Yucatán peninsula and blew eighty percent of Jamaican roofs down. One year later, in September, 1989, Hurricane Hugo pounded the northeastern Caribbean, leaving all of Montserrat's 12,000 homeless and pushing the small British island back to the "kerosene age." Hugo also devastated Puerto Rico and obliterated St. Croix's infrastructure and economy to the point that US troops were called from the mainland to restore civil order there (McFadden 1989).

As far as everyday precipitation is concerned, local topography exerts as much control over the actual pattern of rainfall on a given Caribbean island as do prevailing winds and seasonal changes. The eastern or northeastern (windward) sides of mountainous islands are usually rainy (commonly over 100 inches per annum) and windswept while the southwesterly (leeward) sides are decidedly drier. The startling contrast between the verdant, northern side of Puerto Rico and the semi-arid southern half of the island, for instance, is noticeable to anyone traveling via the high-speed highway that bisects the center of the island. When rain-bearing northeasterly tradewinds abut the highest mountains of the Greater Antilles, massive cumulous cloud formations tower above the islands such as over Hispaniola's Cordillera Central during late summer afternoons.

Even the much-publicized image of the Caribbean region as a homogeneous paradise of tropical warmth must give way to differentiation produced by local topography. Residents of Port of Spain, Trinidad, swelter during the summer months as the mountains north of the city block what little wind blows. Affluent Trinidadians drive north ten miles, through the mountains, to enjoy a cooling sea breeze. Much the same is true of the other islands of the Lesser Antilles whose capital towns all were originally established on the protected leeward sides of the islands during the days of sailing vessels. The northern coasts of Cuba and Jamaica, normally cooler and rainier than the southern parts of the islands, are occasionally susceptible to pulsations of relatively cold air from North America, weather events that preclude the cultivation of temperature-sensitive crops such as cacao (Watts 1987: 15).

The low-lying Caribbean islands in the outer arc of the Lesser Antilles are drought-prone. Little topographic relief triggers little rainfall, and notably Antigua but also Barbados have histories of crop desiccation, drought-induced out-migrations, and serious water shortages. But the volcanic islands immediately to the west also have experienced prolonged periods of drought, and houses throughout the small islands have their own water cisterns. Aruba, Bonaire, and Curaçao are among the driest of all the islands, aridity related both to negligible relief on any of the three as well as to a complex diverging (downward-moving) wind system that contributes to the anomalous aridity on the northern coastal fringe of neighboring Venezuela.

Drought conditions (too often considered simply "natural" weather events) in the Caribbean region are almost certainly more prevalent in the late twentieth century than they were five hundred years ago. From all accounts, Christopher Columbus encountered forested islands whose protective vegetation helped to maintain soil moisture, thereby representing a buffer against seasonal aridity or lasting dry spells. Soil erosion has thus not been the only result of the subsequent deforestation of the Caribbean.

## Pre-Columbian ecology

When the first people arrived in the Caribbean, the islands were more – probably much more – heavily forested than now. In the mountainous areas of the Greater Antilles and on the windward slopes of the smaller islands, true rainforests featured protective leaf canopies and an array of creepers and vines. At intermediate elevations, seasonal subtropical forests were lush during summer and noticeably thinner during the November-to-May dry season when the trees dropped many of their leaves. A xerophytic complex of cacti and thorn scrub created brambles and thickets in the lowest elevations on many of the islands. Vegetational complexity, represented by the number of species on a given island, increased with island size (Howard 1973).

The native fauna was similarly rich. The earliest Spanish visitors to the region recorded an abundance of parrots, pigeons, and doves, available almost for the taking. The richest environmental zone was the shallow sea water surrounding each island where "shellfish, fish, turtles, marine mammals, and waterfowl" abounded (Sauer 1966:

58). Manatees and green sea turtles probably constituted a major source of protein for the island Arawaks (Watts 1987: 61–62). Pre-Columbian hunters of Trinidad subsisted in part by hunting small deer and peccaries, although neither of these continental animals was found farther north in the islands (Newson 1976: 23).

Aboriginal peoples inhabited the Caribbean region for several thousand years prior to Columbus's arrival, but they disappeared rapidly thereafter in the face of enslavement and introduced disease. In the late twentieth century only a few Carib Indians – with a considerable black admixture – are found in isolated reserve areas of Dominica and a few on St. Vincent. Conventional descriptions of Caribbean aboriginal "contributions" to the present are thus usually limited to the identification of a few cultivated plants and the recitation of a handful of contemporary words such as canoe, hammock, and hurricane. (The term "Caribbean" itself is derived from the Caribs who inhabited the Lesser Antilles when the first Europeans arrived.) Yet linkages between the region's aboriginal past and the present may be stronger than most of us think. Aboriginal peoples of the region and imported slaves did overlap, however briefly. On French Martinique, Carib fishermen passed on local fishing techniques to African slaves, and elements of a Pre-Columbian fishing culture have endured there until the present day (Price 1966). Similarly, first African slaves on Hispaniola adopted both cultivation techniques and some New World subsistence crops from the local Arawaks before the latter group was obliterated (Sauer 1966: 211–12).

The earliest human entry into the Caribbean was perhaps as far back as 5,000 BC. Archeological evidence from the southwestern coast of Hispaniola suggests that the earliest inhabitants manufactured flint spear points used in hunting manatees as well as ground animals. Similar artifacts from Belize reinforce the hypothesis that first humans may have come to the Caribbean from Central America (Rouse 1986: 129–33). These early people may have sailed or drifted to the Greater Antilles from Belize, Honduras, or Nicaragua. Whatever their geographic origin, this earliest group of Antillean peoples were augmented much later by others who sailed to the Greater Antilles from Venezuela or the Guianas at about 1,000 BC, apparently bypassing the Lesser Antilles after colonizing Trinidad (Watts 1987: 45–48).

When the first Spaniards arrived in the Greater Antilles, a small number of the earliest inhabitants of the West Indies still lived in an isolated coastal zone of western Cuba and in the interior of

*Osteoautofora c*

Hispaniola (Sauer 1966: 48). Variously referred to as "Ciboney" or "Guanahacabibe" in Cuba, they subsisted strictly on what they hunted or collected: fish, shellfish, turtles, iguana, manatees, birds, or various gathered plants. Their material goods included wooden and stone implements and dug-out canoes. Compared to the aboriginal peoples who followed them, these earliest Caribbean peoples had only crude habitats, dwelling places located under ledges or in coastal caves. The Pre-Columbian peoples who supplanted the earliest "Ciboney" (not a tribal name but one that designated a subservient class) in the Greater Antilles were the Arawaks and Caribs. Linguistic and archeological evidence agree that these two latter groups entered the region as part of a single population migration at about the time of Christ (Rouse 1986: 153). Speaking very generally, the Arawaks inhabited the Greater Antilles and the Bahamas when the Spaniards came, while the Caribs were in the Lesser Antilles. Both groups cultivated a variety of root crops and seed plants, although the Arawaks were considered better farmers, and Carib subsistence was focused on fishing. Carib raiding parties attacked Arawak settlements, carrying off women. The distinctions between Arawaks and Caribs were apparently blurred by Spanish priest/ethnographers who considered friendly Indians "Arawaks" and warlike peoples "Carib," with the result that the extent and influence of the latter group may have been exaggerated (Newson 1976: 17–18).

The most culturally sophisticated and populous island in the Pre-Columbian Caribbean was Hispaniola where the Taino (meaning "good" or "noble") Arawaks had developed a society comparable to the early neolithic cultures of Europe (Moya Pons 1984: 20). The Arawaks recognized a rough partitioning of Hispaniola (which they called Quisqueya) into five provinces, although there was no rigid aboriginal political hierarchy such as those the Spaniards subsequently encountered in Mexico or Peru. Arawak village settlements contained up to several hundred loosely clustered wood-and-palm thatch dwellings, the largest house occupied by a local chief or cacique. The dwellings ringed a central dancing area or meeting ground of beaten earth where villagers also held ballgames on occasion (Sauer 1966: 63–64).

Much more than the early hunter-gatherer peoples of the Caribbean, the Arawaks and Caribs modified the insular environments in the centuries before Columbus. Aboriginal burning likely extended zones of savanna grasslands at the expense of forest in the Greater Antilles (Watts 1987: 82). Some of the "natural" vegetation of the islands was apparently introduced through aboriginal migrations by either accident or design (Howard 1973: 16). Environmental modification was, furthermore, intensified by a possibly brisk aboriginal trade network dealing in shells, manufactured stone and wood objects, salted fish, whale oil, containers, ornaments, game animals, and vegetable products, a trade that ranged from Trinidad and the Guianas through the Lesser Antilles and possibly as far north as Puerto Rico (Newson 1976: 61–62).

Pre-Columbian agricultural systems in the Caribbean were complex, apparently prolific, and – supplemented with fishing and hunting – provided the great bulk of foodstuffs (Watts 1987: 53–60). Cultivators cleared their forest plots ("conucos") adjacent to the village settlements, burned the cuttings, mounded the cleared earth into small hillocks, and then planted a variety of foodcrops therein. The main cultivated plants were the root crops of manioc (from which cassava bread was produced) and sweet potatoes, although other root crops, including peanuts, and also corn and beans were grown. Caribbean "conuco" agriculture, like agricultural subsistence production throughout the tropics that is variously termed "shifting" or "slash and burn," mimicked the species' diversity and the vertical structure of the surrounding natural vegetation. Also, as in subsistence agriculture elsewhere, aboriginal cultivation in the Caribbean seems to have involved all family members for clearing, planting, weeding, and harvest.

The Caribs kept a small duck, a domesticate whose distribution did not extend into the Greater Antilles. Aboriginal people throughout the region had a kind of voiceless dog which doubled as both a house pet and a food item. A type of woven cotton provided material for women's skirts. And a diverse array of fiber nets, including both hammocks and fishing gear, augmented the material culture of Pre-Columbian peoples of the Caribbean (Sauer 1966: 59–61). But a lifeless inventory of economic activities and related material possessions is only a caricature of the people themselves, a point especially relevant for the aboriginal peoples of the Caribbean because they disappeared so rapidly when Europeans arrived.

Both Caribs and Arawaks seem to have had complex religious beliefs. The notion of travel to an after-life is supported from funeral offerings of water, food, and implements found in Trinidad grave sites (Newson 1976: 65–66). Icons of wood, stone, and bone in the Greater Antilles were associated with Arawakan religious spirits or *zemis*. Arawak myths and folktales were associated with religious expression and bound up with their own identities within their insular environments. One Arawakan folk tale involved group formation and the importance of responsible leadership, a tale possibly related to the relatively recent entry of Caribs and Arawaks into the Caribbean region from northern South America (Stevens-Arroyo 1984).

How many people lived in the Caribbean region prior to the coming of the Europeans? Estimates vary widely. The most common early estimates by the Spaniards was slightly over one million for Hispaniola at the time of contact. Later estimates, based on a combination of archival reports and a calculated environmental carrying capacity for agriculture, range on the high end to as many as six million for all of the islands together (Watts 1987: 74–75). The estimates by most historians are more conservative, such as the one offered by Franklin Knight who suggests that the indigenous populace of the entire Caribbean in 1492 was no larger than "three-quarters of a million, the great majority of whom lived on . . . Hispaniola" (1978: 5).

Probably a consensus aboriginal population estimate will never be possible owing to the speed with which the aboriginal West Indians disappeared after European contact. The first census on Hispaniola in 1508 (which was probably little better than a population estimate) counted only 60,000 Arawaks left. After smallpox swept the island in 1519, an incredible 2,500 remained (Moya Pons 1984: 47). The demise of the aboriginal peoples of the Greater Antilles was the first in a series of massive environmental transformations which began in the autumn of 1492. Christopher Columbus died thinking that he had discovered a series of islands off the east coast of China, but he had actually established Europe's first overseas colony.

### Spanish transformations

Columbus's voyage of discovery also represented a spatial expansion of Europe's economic and political control (Wolf 1982: 108–25). Spanish mercantilism in the years before the voyage was facing reduced surpluses in part because of competition with Portuguese and Genoese merchants. Portugal, moreover, had gradually extended its exploration and trade network down Africa's west coast. This extension was part of an overall plan to circumvent the Muslim middlemen who intervened geographically and thereby controlled the increasing flow of goods from the Orient to Europe. So Columbus's well-known strategy of sailing west in order to go east was a Spanish countermove. His sailing route from Spain was facilitated by the easterly tradewinds off the Azores high, winds well known to Columbus and other veteran Mediterranean sea captains. In his initial trans-Atlantic voyage, Columbus steered a northerly direction across the trades in order to reach the presumed latitude of Cathay (Watts 1987: 87). The return journey was north and then east to Europe via the westerlies, thereby establishing the return sailing route between the Old World and the New that European seafarers would follow for centuries thereafter.

The Arawaks greeted Columbus and his party with apparent warmth, and Columbus transported several of the aboriginal peoples home with him on the return voyage along with a few animals, cultivated plants, and selected native implements. Before his return from the Caribbean, Columbus sited the small encampment of Navidad ("Christmas") on Hispaniola's north coast and left a few Spaniards behind. When Columbus returned to Navidad in late November, 1493, all of the Europeans had died, probably from syphillis they had contracted from the Arawaks. If syphillis were the reason, it was an aboriginal-to-European disease transfer that would be reciprocated many times over by the Spaniards who subsequently transported a complex of virulent new diseases into the Americas (Crosby 1972).

Among the objects he carried back to Spain, Columbus took gold nuggets from an inland placer in Hispaniola. An active Spanish search for gold followed in the Greater Antilles, mainly in Hispaniola, during the next three decades. This quest for gold, furthermore, was directly related to the plummeting population of the aboriginal peoples of both the Greater Antilles and the Caribbean region as a whole. The Spaniards herded enslaved Arawak men of Hispaniola into the small mines of the island. The crowding heightened the transfer of disease among individuals, and the removal of cultivators from their agricultural plots created food shortages that further weakened the laborers and increased their disease susceptibility. In response to the accelerating death rate on Hispaniola, Spanish slave-raiding parties tapped the aboriginal populations of adjacent islands, notably Jamaica and Puerto Rico. Spanish slavers also brought an estimated 40,000 Arawaks from the Lucayas (Bahamas) to work the mines of Hispaniola in the first decade of the sixteenth century. In 1511 they had ranged as far south as Trinidad where enslavement practices eventually led to similar famine and human population decline (Newson 1976: 72, 77–78). The "weakness" of the aboriginal Caribbean peoples in the first few years of contact – a curiously benign term for what seems actually to have taken place – also helped to inspire the importation of the first African slaves into the Caribbean. A few black slaves, who appeared notably stronger than the Arawaks, had been brought to Hispaniola, perhaps before 1500. Ferdinand of Spain referred to the importance of a shipment of seventeen black African slaves who arrived in Hispaniola in 1505 because "all of these be getting gold for me" (Sauer 1966: 206–7).

Spanish activities in the Caribbean in these earliest years of colonization did not always stem from a monolithic economic strategy. The Crown acknowledged an obligation to protect the freedom of peaceful Caribbean Indians (although slave-raiding parties often ignored edicts against enslavement.) And the well-known writings of the Dominican friar Bartolomé de Las Casas, who traveled through the region and chronicled the terror and tragedy of the "leyenda negra" (black legend) perpetrated against the Indians by local Spaniards, served to highlight and also to intensify the rivalries between European Spain and New World Spaniards. In 1503 the Crown invested the ultimate authority for Caribbean policy and economic matters in the *Casa de Contratación* located in Seville, an institution that decreed a virtual Spanish monopoly over the Caribbean. In this way, Spain attempted to ensure that profits from Caribbean minerals, hides, agricultural products, work and all else would accrue to the mother country.

The leather hides started coming from Hispaniola in the first years of the sixteenth century; by the 1520s "many herds" of cattle existed on the island, some reported to be as large as eight thousand (Crosby 1972: 87). Those kinds of reports must have astounded Spaniards staying behind; livestock-keeping had long been practiced in Iberia, but Spanish stockraising at the time consisted of small, mixed herds (Butzer 1988). It was all the more remarkable because cattle – along with horses, pigs, sheep, and goats – had not been introduced into the Caribbean from Spain until Columbus's second voyage (Watts 1987: 90). Facing few predators in the Caribbean, those animals – especially cattle and pigs – multiplied rapidly, many becoming wild. In a broad sense, introduced livestock competed with aboriginal peoples by trampling formerly undisturbed soils and disrupting conuco cultivation mounds. From their new animal breeding grounds in the Greater Antilles, Spaniards introduced European (now Caribbean) animals to other places. The first horses ever introduced to Trinidad, for example, came from Puerto Rico in 1531 (Newson 1976: 88).

The explosion in numbers of cattle over the subtropical grasslands of the Greater Antilles provided a food source for non-Spanish pirates, whose presence in the region presented formidable competition to Spanish colonialism by the seventeenth century. These seafarers of mixed origin are said to have cooked their meat on iron grills (*boucans* in the Arawak language) and were therefore known as *boucaniers* in French, a term subsequently anglicized to "buccaneers." The plunder of the English, French, and Dutch pirates then became active trade and, eventually, colonialism in the eastern Caribbean (Wallerstein 1980: 161). This Northern European foothold had thus been underpinned by the ecological transformations in the Greater Antilles set in motion by the sudden presence of thousands of introduced cattle, transformations that created plenty for pirates and famine for others.

Because of the climate differences between the Mediterranean and Caribbean regions, Spanish agricultural staples did poorly in the latter area. Wheat did not fare well in the Caribbean humidity, and grapes and olives needed the pronounced aridity of Mediterranean summers. The Spanish then demanded food tributes from the Arawaks to compensate for subsistence shortfalls. And the ensuing famine periods that were exacerbated by Spanish colonial practices affected the oppressors as well as the oppressed: the Spaniards always eagerly awaited supply ships from home in the early years of Caribbean colonialism.

Columbus brought sugar cane – the crop that has become nearly synonymous with the Caribbean – to Hispaniola on his second voyage; once planted, the cane germinated in seven days (Deerr 1949: 116–17). On Hispaniola the planting and processing of sugar cane achieved limited success in the first three decades of Spanish occupation. The principal growing area was west of Santo Domingo on Hispaniola's south coast. The deforestation of the land to make way for cane planting was accomplished mainly by black slaves (Sauer 1966: 210). Removal of the natural vegetation also produced wood for fuel, lumber for housing, fences, and corrals, and it represented the first example of the massive environmental change that would transform the region in the seventeenth century.

Enthusiasm about the Caribbean region among the New World Spaniards waned by 1520 as Hernán Cortés located and plundered the aboriginal mainland civilizations in Mexico, and Francisco Pizarro conquered Peru quickly thereafter. The Spanish Crown nevertheless continued to regard the islands as locationally strategic. Fortresses in the Greater Antilles guarded the incoming, and especially the return, voyages of the annual treasure fleets from Cartagena and Vera Cruz (West and Augelli 1976: 63–64). These flotillas, bound for the warehouses of Seville, were, as the sixteenth century wore on, subjected to increasing attack by the pirates from European nation states that were beginning to rival Spain for parts of the region.

In order to assure a Spanish presence in the islands, the Crown imposed severe punitive measures against those Spanish colonists attempting to emigrate from the Antilles to the mainland. And reluctant Spanish officials residing in the Greater Antilles, after the smallpox epidemic of 1519 had left only a few sickened Indians behind, could see inscribed upon the landscape the environmental results of only three decades of Spanish control. Abandoned Arawak village areas, whose beaten earth had recently been the scene of dance and play, were now rotted huts surrounding weedy growth. Cleared conuco plots, formerly productive entanglements of root and garden crops, were now stands of sugar cane tended by African slaves. Savanna grasslands and adjacent forested areas, formerly devoid of grazing animals, now supported horses, cattle, and pigs, whose trails and trampling increased soil wastage. Deforestation produced similar erosion. The prospect was not altogether pleasing. The late geographer Carl Sauer provided a succinct environmental assessment of the Caribbean islands and rimlands after fewer than three decades of European control: "By 1519 the Spanish Main was a sorry shell" (1966: 294).

### The great clearing

Although formally claiming the entire Caribbean, Spain never actively colonized the Lesser Antilles except for her occasional slave-raiding forays and a modest presence in Trinidad. English and French seafarers, by the early seventeenth century, were familiar enough with the small islands to put ashore occasionally in order to take on fresh water and hunt food. Active northern European settlement of the eastern Caribbean did not come until 1624 when small contingents of French and English each arrived in St. Kitts, an island they inhabited jointly for nearly a century. The English then occupied uninhabited Barbados in 1627, and the French settled Guadeloupe in 1635. These tiny settlements represented the beginnings of an Anglo-French domination of the eastern Caribbean that has lasted – formally for France because Guadeloupe, Martinique, and French Guiana are still departments of metropolitan France – for over three and one-half centuries.

Relationships among the earliest French and English settlers in the region, especially within particular islands, were often hostile. On St. Kitts one of their few collaborative activities was a joint sneak attack to eliminate the local Caribs (Dunn 1972: 18). They also joined forces in attempting to repel occasional Spanish attacks, although by the late seventeenth century Spain had surrendered Jamaica to the English (1670) and acknowledged English sovereignty over several of the smaller islands.

For the first three decades, seventeenth-century European settlements of the Lesser Antilles were based upon modest subsistence and cash-crop cultivation by yeoman farmers, habitation patterns not unlike those in eastern North America at the time. In the islands, farmers from the northwest of France and England's southern counties planted tobacco, indigo, and both Old World and New World subsistence crops in a strange new environment: a broiling tropical sun evaporated standing water and baked soil at an unbelievable rate; yet a heavy afternoon thunderstorm could create near-flooding conditions on the same day. Clearing farm plots from the natural vegetation along island littorals, moreover, was arduous, accomplished usually by "ring-barking" and burning, much as the Caribs had done before (Watts 1987: 154; Innes 1970).

Then sugar cane was introduced to Barbados in about 1640, not by the Spanish via the Greater Antilles but by the Dutch via northeastern Brazil (Galloway 1989: 77–83). Cane cultivation thereafter spread rapidly to St. Kitts, Nevis, and Antigua by the 1650s and to the other small islands, including those controlled by France, shortly thereafter. An exact tracing of the diffusion of sugar cane throughout the Lesser Antilles is, however, much less important than it is to emphasize the rapid, sugar-based ascendance of the English and French islands from minor semi-subsistence colonial outliers to vital sources of national and individual wealth. English, French, and Dutch vessels soon vied to carry precious cargoes of semi-refined Caribbean sugar to Europe's growing market. And sugar production, because of European market demand, then began to imprint the Caribbean in fundamental ways.

The ecological impact of the colonial Caribbean sugar plantation era has received less academic attention than has its commercial, social, and demographic dimensions. As sugar cane cultivation on the small islands outcompeted all other crops and demanded more and more acreage, forest clearance coincided with the massive introduction of West African slaves who then accomplished the large-scale clearing. Under European supervision, they removed the forest cover from insular soils after girdling and burning the trees.

By 1665 only isolated pockets of forest remained in the few inaccessible zones of Barbados (Watts 1987: 219). Removal of the forest cover occurred shortly thereafter in the volcanic, and therefore more rugged, Leewards. On island after island, the reasonably flat land was cleared for canes in an astonishingly short period, probably by about 1680 for Barbados and the English Leewards. In the quarter century since sugar cane had arrived in the eastern Caribbean, the sounds of axes and smells of woodsmoke must have been universal in the small islands, a brief period referred to by historians Carl and Roberta Bridenbaugh as "The Great Clearing" (1972: 268).

The abruptness of this change was unprecedented, and it represented a sharp ecological discontinuity with the past. Local reports of soil erosion and similar environmental stress occurred almost simultaneously with the clearing, but Caribbean environmental decisions were no longer being made in the Caribbean itself. The islands suddenly had been absorbed into an expanding European-centered commodity exchange of trans-Atlantic scope. And growing European market demand increased sugar productivity schedules that knew or cared little about insular soil erosion rates or the heightened drought susceptibility that deforestation created. The latter problem occurred in the English Leewards where slaves felled trees with axes, burned the cleared brush, and then planted young canes amidst the ashes (Harris 1965). Sensational cases of Caribbean soil erosion, especially from mountainous islands, were relayed to metropolitan governments during the subsequent century; on volcanic Nevis, for example, heavy rains created so much sheet erosion from unprotected higher slopes as to enrich temporarily the slopes below (Coke 1811: 4–5). Perhaps the most spectacular case of environmental devastation during these devastating decades in Caribbean environmental history was the French burning of the forest and scrub cover of the entire island of St. Croix "To remedy the inconvenience and make the island more healthy" (Dirks 1987: 16).

Land-use transformation on the French islands of the eastern Caribbean followed a course parallel to that of the English possessions. The early French policy of establishing a transplanted European peasantry on Guadeloupe and Martinique was soon overridden. By the late seventeenth century sugar plantations had begun to dominate the flatlands of Martinique, and thousands of African slaves had been introduced. Clearing of the natural vegetation on Martinique for cane fields led to similar environmental deterioration as in Barbados and elsewhere. By the early eighteenth century in Martinique, land-use decisions were being made exclusively in France, and "[a]s a consequence of severe erosion, the moisture-retaining capacity of soil in different regions (of Martinique) was reduced, and plants in those places were subjected to greater drought stress than the rainfall warranted" (Kimber 1988: 171, 180–81). St. Domingue (later Haiti) was not universally recognized as French soil until 1697, but by 1701 in St. Domingue "there were already 35 mills at work grinding cane" (Deerr 1949: 239), and clearing of vast portions of the western half of Hispaniola was beginning to set in motion destabilizing ecological processes that ultimately would lead to some of the most severe environmental problems in the world.

Plantation demands on insular ecosystems went far beyond the necessity for cleared, and therefore deforested, land. Timber also was necessary for buildings, fences, and animal pens. Boilers needed fuel to reduce sugar juice to granular sugar for export. Coal sometimes came from England as ship ballast, although the constant need for fuel led to the rapid deforestation of steep slopes on Caribbean islands themselves, and lumbering expeditions occasionally visited small, non-plantation islands in order to cut firewood (Sheridan 1973: 115). The overall environmental deterioration on some of the smaller plantation islands led officials to seek solutions within the context of the overall plantation system itself. Planters on English Barbados began to substitute dried cane stalks (bagasse) for boiler fuel before the end of the seventeenth century (Watts 1987: 398–99). And the possible application of essentially unlimited quantities of slave labor could, at least in the minds of some colonial officials, compensate for environmental decline. In 1734, Governor William Mathew of the British Leeward Islands called for a higher slave density so that a more intense labor routine might substitute for soil quality deterioration on St. Kitts (Richardson 1983: 56–57).

Although environmental deterioration accompanied the plantation "development" of these small islands with punishing long-term results, it is important to understand that the islands were not simply degraded but also transformed. A host of new cultivated plants and animals were introduced so that the islands might produce sugar cane and other tropical staples more efficiently. The introduction of Iberian cattle was only an early and dramatic example of this transformation that would involve dozens of new plants and animals brought to the Caribbean over a period of decades, introductions that were both planned and accidental. A famous example of the conscious introduction of food plants to the Caribbean to improve the local subsistence base was Captain William Bligh's bringing the breadfruit to St. Vincent and then Jamaica from the South Pacific late in the eighteenth century (Howard 1973: 16).

The environmental transformation accomplished by African slaves along the coastal mudflats of the Guianas came somewhat later, although the changes in the Guianas were perhaps more dramatic and, for the slaves, certainly more arduous. The English had planted sugar cane in Suriname which they controlled until it was ceded to Holland in 1667. The Dutch, conversely, claimed Guyana until the three colonies of Essequibo, Demerara, and Berbice became British in 1814. As the soil fertility declined in the tiny islands to the north and as the average size of sugar-cane plantations enlarged in response to technical changes, the fertile alluvium of the Guianese coastal plain became ever more attractive to European planters. But reclaiming coastal estates from tidal seaflats required more than simply clearing and then planting in the openings. Not only did mangrove and swamp forest have to be removed, but protective seawalls were obvious requirements, as were drainage and irrigation canals, before sugar cane or other tropical staples could be planted. As Guyana passed into British hands in the early nineteenth century, estate owners extended reclamation efforts – formerly limited to riverine locations – along the ocean littoral. Thousands of slaves then transformed the sea-level mudflats into a checkerboard plantation landscape criss-crossed with dikes, dams, and canals. The late Guyanese historian, Walter Rodney, estimates that this massive environmental change "meant that slaves moved 100 million tons of heavy, waterlogged clay with shovel in hand, while enduring conditions of perpetual mud and water" (1981: 3).

Slave emancipation in the British Caribbean in the 1830s, followed by French, Dutch, and Danish slave emancipation in the following decade, was accompanied by technical changes with ultimate environmental consequences. In adjusting to the loss of a captive labor force, British and French planters in the Caribbean experimented with a variety of new agricultural techniques throughout the latter part of the nineteenth century. The increasing use of the steam engine to grind canes heightened the throughput capacity of individual factories and thereby enlarged the sugar-cane estates tributary to each mill. A new kind of factory, the central, ground canes from entire island districts, and did it more thoroughly than before. Central sugar factories appeared in Martinique and Guadeloupe in 1847 and in the British Caribbean three decades later. The Usine St. Madeleine, an enormous sugar central, was constructed in 1872 in southwestern Trinidad, helping to explain the rapid removal of the rainforest for canelands on the rich zone of alluvial soil in the western part of the island (Wood 1968: 295–97).

The Cuban sugar industry, dormant during the plantation heyday of the Lesser Antilles, had begun to stir by the 1800s. The overwhelming majority of nineteenth-century Cuban sugar subsequently came from the central and western parts of the island, the areas immediately adjacent to Havana. And the increasing need for open land and boiler fuel led to the same kind of forest removal in Cuba that had occurred nearly two centuries earlier in the small islands to the east and south. In nineteenth-century Cuba, tree clearing and burning was accomplished principally by a free peasantry because slaves (final emancipation in Cuba was not until 1886) assigned to forest removal often escaped into the woods. Cuban historian Manuel Moreno Fraginals has written a vivid and moving summary of what he calls "The Death of the Forest" in nineteenth-century Cuba. In the three centuries since Columbus, the Cuban forests had been known universally as a source of wood of remarkable quality, and palaces and cathedrals in both Europe and the Indies abounded with doors, windows, and artifacts crafted from Cuban hardwoods. But the tradition and wealth of the great Cuban forests were swept away by sugar in the central and western parts of the island: "Sugar exterminated the forests. Deaf and blind to history, focusing on the present, the sugarocracy destroyed in years what only centuries could replace – and at the same time destroyed much of the island's fertility by soil erosion and the drying-up of thousands of streams" (Moreno Fraginals 1976: 76–77).

As the nineteenth century became the twentieth, and as Cuba's status as a political colony of Spain changed to that of an economic colony of the United States, American influence helped to complete the deforestation of most of Cuba. By 1916 the Cuban Railroad, inspired by American financial speculation in Cuban sugar, tapped the heretofore inaccessible upland areas of eastern Cuba: "Here were vast forests, to be purchased cheaply, growing on cane-land of unequalled richness ... the forest received scant shrift. Corps of wood cutters were set to levelling trees. Where there was hardwood, it was sometimes hauled out by ox and chain. The remainder of the trees lay as they fell for several months, and then were fired – thousands of acres at a time, in a conflagration that drew the entire countryside to prevent its spreading. Cane was planted between the blackened stumps, without the trouble of plowing" (Jenks 1928: 181).

American influence was similar in Hispaniola and Puerto Rico early in the twentieth century, with similar environmental consequences. A branch of the American-owned South Puerto Rico Sugar Company established the La Romana sugar estate on the southern coast of the Dominican Republic in 1911. It was the largest among fourteen American-controlled sugar plantations in the country. By 1925 the American estates covered an incredible 366,000 acres (over 570 square miles) of sugar cane that had been carved out of the subtropical forests and guinea grass of the lightly settled southeastern quadrant of Hispaniola (Knight 1928: 139). The annexation of Puerto Rico as an American Commonwealth in 1898 led to similarly unprecedented growth of the sugar-cane and tobacco-growing industries in that easternmost island of the Greater Antilles. By 1912, along the southern coast of Puerto Rico, a new railroad had been built to expedite the transportation of sugar cane. Small-scale peasants in the region found it nearly impossible to find firewood that they previously had gathered "as the coastal woodland was rapidly cleared, the logs being used as railroad ties" (Mintz 1953a: 245).

The Greater Antilles was not the only focus of United States commercial interests in the circum-Caribbean realm at the turn of the century. The Boston Fruit Company, whose banana plantations established in Cuba and the Dominican Republic in the 1880s had been converted to sugar lands, subsequently sought previously undeveloped and hurricane-free lands along the Caribbean rim of Central America. A corporate amalgamation transformed the Boston Fruit Company into the United Fruit Company in 1899. And, during the next decade, the company's American supervisors and Jamaican laborers cleared thousands of acres of coastal scrub and rainforest mainly in Honduras and Costa Rica. By 1914 the sense of pride and technical omnipotence surrounding the American environmental conquest of coastal Central America seemed boundless: "An empire of agriculture was carved from the jealous and resentful jungles. Hundreds of miles of railroads were constructed into wilderness . . . from out of the waters of the Caribbean steamed scores of ships to the marts of the Old and New World bearing the commerce which Yankee enterprise had created in a crusade to attain the peaceful Conquest of the Tropics" (Adams 1914: 122).

### The geographical legacy

Sugar cane remains the principal crop of the Greater Antilles and also the Caribbean as a whole in the late twentieth century. Control over the cultivation of much of the region's cane has come under local government auspices or at least local government supervision. Much has changed in the technology and scale of cultivation and also in processing techniques during the five hundred years since Columbus brought cane cuttings to Hispaniola. Yet the vast majority of Caribbean sugar still is sold in North America and Europe, and the product is thus subject to external fluctuations in sugar demand and sugar prices. So the cultivation of Caribbean sugar cane, although obviously much different from when it was first introduced to the region, remains an emblem of outside economic influence over the region.

The colonial clearing of insular forests for sugar cane, although cane now has disappeared in many of these places, also set in motion a chain of ecological events that is vitally relevant in understanding the long-term degradation of the Caribbean environment. During slavery, the production of local subsistence crops was limited and subject to direct planter supervision, an issue discussed further in the following chapter. Emancipation then had important environmental implications, as well as readjusting social relationships in the region. Throughout the Caribbean after slavery, "reconstituted peasantries" (Mintz 1974b) composed of former slaves sought their own subsistence and their own identities by establishing themselves on the land. Plantations, in general, continued to preempt the region's best lands, so communities of freedmen commonly were relegated to marginal environmental areas – usually hillsides but sometimes swampy lowlands – that often had been considered unsuitable for estate cultivation (Richardson 1984).

These changes had profound environmental effects in the region for at least three reasons. First, the (heretofore captive) labor force of the Caribbean, whose foodstuffs often had been imported, now relied almost exclusively on the local environments for sustenance, adding additional subsistence pressure. Second, the marginal hillside character of the new village settlements often led to severe cases of sheet erosion. Third, former slaves usually avoided planting sugar cane, preferring instead crops that they and their families could eat; and the abandonment of sugar cane (which, as a grass, has a beneficial side-effect of anchoring the soil) in favor of clean-row tillage, led to further erosion.

Small-scale subsistence agriculture in the Caribbean region has continued to expand since emancipation at the expense of former sugar-cane acreage, especially in the smaller islands. It is a principal means by which the descendants of an introduced overpopulation –

whose numbers bore little relationship to the ability of the local environments to sustain them in the first place – can survive. And the considerable subsistence skills of Caribbean peoples notwithstanding, the cumulative soil erosion and ongoing subsistence pressures continue to lead to further environmental degradation.

Livestock-keeping, popular among Afro-Caribbean slaves who used animals as a slender means of establishing "growth capital" in the interstices of a plantation-controlled environment, has nearly taken over islands whose erosion has rendered them useless for anything else. In the tiny Grenadines between St. Vincent and Grenada, or in the drought-prone Leewards, livestock now browses through thorn scrubs where subtropical forests once stood. Animals are ready sources of cash and represent "resources on the hoof" (Berleant-Schiller 1977: 300) as on Barbuda, Nevis, or Anguilla. They also have helped create moonscapes on parts of the small islands whose carrying capacities have been systematically reduced through a series of land-use stages originally set in motion by the plantation clearing that took place centuries ago.

The practical importance of the degraded physical environments found throughout the Caribbean region cannot be overemphasized. Food deficits (discussed further in Chapter 5) are common, especially in the smallest islands. Caribbean island governments acknowledge their needs for a greater independence in food production, citing the region's high populations and high food-importation rates. And there is no shortage of internationally financed "development" reports regarding Caribbean food problems, nearly every one of them prefaced with the immediacy that a local food/population imbalance is dangerous. Yet few of these reports acknowledge fully the underlying reasons for Caribbean food deficits which include a centuries-old preoccupation with cash crops, an introduced overpopulation that began during slavery, and a degraded physical environment whose impoverishment can be traced to colonial control.

The Caribbean was Europe's first overseas colony. Yet, except for the obvious familiarization with the tropics that the Caribbean provided many Europeans, one cannot assert that Caribbean colonization techniques were spread elsewhere, as from a New World "training ground." This is because the Caribbean also has been unique in colonial history: the aboriginal peoples of the region were obliterated almost upon contact. A "native" African work force, "guests" rather than "hosts" (Mintz 1974b), then was imported. Also, the region's insularity allowed colonial masters almost complete domination of the total environment, different from larger, mainland areas colonized later where native peoples were able to maintain indigenous subsistence systems. Even in nearby Yucatan, local aboriginal control over portions of the local landscapes provided both physical subsistence as well as symbolic sustenance, helping to account for the remarkable conservatism and resistance of the Maya to centuries of nominal external control (Clendinnen 1980).

Caribbean peoples have been denied a similar geographical experience owing to the nature of the region's colonial past. The land that they have inherited has for centuries been a vehicle for profit, not husbanded for future generations. As economic surpluses have flowed from the Caribbean region to meet metropolitan market demands, local Caribbean environments accordingly have been degraded. The well-known "man-land" term, an expression used for decades among American cultural geographers to sum up the intimate relationships between small-scale non-Western cultivators and their immediate surroundings, is, in the case of the Caribbean region, curiously inappropriate. People and lands in the Caribbean region during the centuries of colonial control were factors of production, their interactions controlled and directed by plantation supervisors. Only after these centuries have the region's peoples and their lands been truly united and now under difficult circumstances, not the least of which is a legacy of environmental deterioration. Whether the small-scale cultivators of the Caribbean region are able to survive and prosper in light of this ecological inheritance poses a stern test of their considerable imagination and creativity.

# 3

## Plantations and their peoples to 1900

The Caribbean plantation is a global, not a regional, enterprise, and it has always been so. The economic historian Richard Sheridan, in describing the British West Indies in the seventeenth and eighteenth centuries, points out that factors of production from throughout the world were united in Caribbean settings, and the unique sum of these introduced parts resulted in the Caribbean plantation: the original domestication site of the principal cash crop (sugar cane) probably was southern Asia; the labor supply and some food crops were of West African origin; building materials, livestock, and food came from eastern North America; and the capital, managerial expertise, and technology were from western and southern Europe (1973: 107).

The warm climates and virgin soils of the Caribbean islands and rimlands underpinned the production of tropical plantation products for the mid-latitude European market. This production was of modest scope under the Spanish in Hispaniola in the early 1500s, but it began in earnest with the English, French, and Dutch colonization of the Lesser Antilles one century later. The trade circulation of plantation products eventually was interlinked with the movement of global inputs for the Caribbean plantation in a complex shipping network often described as the "triangular trade," involving Europe, West Africa, and Atlantic America from the Guianas to what is now Canada. The classic three-sided trade involving the Caribbean region usually is explained thus: manufactured goods from Europe were sent to West Africa, then slaves transported west to the Caribbean region, and tropical staples shipped from the Caribbean back to Europe, a clockwise sailing circle facilitated by the rotation of the winds blowing out of the subtropical high pressure cell in the north central Atlantic.

Several variations of this trade, however, linked the Caribbean, Europe, Africa, and North America. Molasses from the Caribbean often went to North America, for example, and then rum from North America to West Africa. Sailing ships from North America also hauled lumber and salt fish south to the islands, beating their way against the easterly crosswinds and taking care not to be caught in open waters during the July to October hurricane season. And, although the "triangular trade" terminology is useful and vivid, the importance of the trade was that it represented a commodity flow that tied together far-flung zones in different corners of the Atlantic. The ships themselves – as some unromantic scholars have pointed out – often did little more than alternate back and forth between two areas or ports (e.g., Wallerstein 1980: 238).

The impact on Europe of the Caribbean plantation and its products was momentous. The wealth accrued from the plantation trade affected European politics and trade. The late Eric Williams, an historian who later became Prime Minister of Trinidad and Tobago, has forwarded the most famous thesis to date concerning the overall importance of Caribbean plantation slavery when he asserted that capital accumulated therefrom financed England's commercial and technical advances culminating in the industrial revolution (1944). Williams's thesis continues to attract attention and to inspire provocative reactions (e.g., Solow and Engerman 1987). And contemplating the importance of the Caribbean plantation along interregional lines underscores the many ways in which the institution transformed people and places on both sides of the Atlantic:

The plantation system which extirpated the yeoman farmers of the non-Hispanic Antilles and scourged West Africa for centuries richly rewarded its organizers. The wealth produced by African slaves on land wrested from Arawak and Carib Indians flowed into the European metropolises in great rivers, nourishing infant industry, making possible the foundation of great families, and supporting the growth and spread of culture and civilization in the form of universities, libraries, museums, and symphony orchestras. Plantation products also did their part for Western civilization. West Indian tobacco, coffee, sugar, and rum, together with Indian tea, were effective fare for the factory workers of Britain and France, quelling their hunger pangs and numbing their outrage. (Mintz 1964: xvi–xvii)

Put simply, prior to the nineteenth century the plantation islands of the Caribbean were "the most valued possessions in the overseas imperial world" (Knight 1978: 40). We already have noted how this value was translated into long-term environmental effects for the

Caribbean region owing to short-term economic decisions made in the metropoles. Yet it is important to understand, or at least to suspect, that the Caribbean region was more than a satellite during its centuries of colonization. The Caribbean plantation, in combining field and factory at an early date, was perhaps "industrial" prior to the rise of industry in Europe (Mintz 1985: 50–52). Furthermore, some of the earliest applications anywhere of modern agricultural and crop-processing techniques have been on Caribbean plantations.

But the selective introduction of fertilizers, new crop strains, or modern irrigation techniques, as all students of tropical agricultural development know, does not necessarily spill over into the rest of the countryside. For decades, indeed centuries, Caribbean landscapes have been landscapes of contrast; the gleaming towers and smoke-stacks of modern estate factories have been literally within sight of tiny, impoverished dwellings occupied by plantation workers, a juxtaposition still obvious throughout the region. This geographical contrast, furthermore, possibly provides a clue to understanding colonial policy beyond the Caribbean region. Anthropologist Clifford Geertz has described Dutch colonial policy in Indonesia, for example, as "one long attempt to bring Indonesia's crops into the modern world, but not her people" (1963: 48)

### Plantation fields and factories

The term Caribbean "plantation" is, strictly speaking, an abstraction because it describes a variety of land units that have grown, collapsed, and changed remarkably at different times over the centuries. Its closest Old World antecedent seems to have been the system of sugar-cane production in Portuguese São Tomé, off the coast of Africa, in about 1500. There the Portuguese significantly "abandoned Mediterranean forms of land and labor in favor of large plantations worked by African slaves" (Galloway 1989: 59).

It would be erroneous, however, to suppose that there was a single progenitor of the Caribbean plantation. In the early seventeenth century, more than a century after Christopher Columbus had transferred elements of the Mediterranean plantation west across the Atlantic, the Dutch reintroduced a sugar-cane cultivation system to the English islands in the Lesser Antilles by way of northeastern Brazil. For the next two centuries, until the early 1800s, the small islands of the eastern Caribbean, both British and French, supported the plantation production of sugar cane marked by intensive slave labor. Jamaica was an early outlier of this eastern Caribbean plantation core with Trinidad and the Guianas coming to the forefront somewhat later. Then the Greater Antilles reemerged as the principal plantation islands of the region in the nineteenth century; this reemergence was intensified with United States' capital (Chapter 4) and accompanied by modern agricultural techniques and economies of scale.

The chronological complexity of the Caribbean plantation has been compounded by environmental differences. Low-lying islands such as Barbados, for example, traditionally had more plantation land given over to cash crops than on partly mountainous islands such as Jamaica or Martinique where "waste" areas could support subsistence crops for local plantation labor forces. Cultural differences added more complexity. The plantations of the French West Indies were reinvigorated with the introduction of the famous Otaheite cane from the Pacific via France late in the eighteenth century (Deerr 1949: 19). And a sophisticated financial credit system among British Caribbean planters at the same time possibly was the eventual cause of a higher rate of absentee ownership among the British as opposed to the French (Wallerstein 1980: 169–70).

Despite many differences, however, the Caribbean plantation has displayed several distinct characteristics transcending time and space. First, plantation land has been considered a commodity or factor of production because capital accumulation or monetary profit has been the plantation's overriding goal. Second, control of the plantation has been centralized, exerted either directly by the owner or through his representatives. Third, a substantial investment in equipment, technology, and auxiliary technicians often has been necessary, again with an eye towards profitability. Fourth, the workforce usually has been introduced from elsewhere and coerced into plantation labor, not always through slavery but also by more subtle means. Fifth, the plantation product has always been produced for an impersonal overseas market. Sixth, and most important, the plantation has been a product of metropolitan capital and enterprise integrated into a Caribbean setting (Beckford 1972: 11; Wolf and Mintz 1957). The economic and cultural characteristics of individual plantations, it is important to note, affect the larger region and, given the characteristic small scale of most Caribbean locales, often influence entire islands (Wagley 1957).

This influence is clear in any assessment of the settlement geography, especially in the past but also in the present, of the Caribbean region. The plantation, not the town or village, was the basic socioeconomic unit of the colonial countryside in the Caribbean.

Saturday morning markets were – and still are in some cases – held on plantation grounds. The plantation, not market towns, was the locus of wages and medical care. Even the term "town" was rarely used in the region, and it continues to be a curiously inappropriate term to apply to small rural Caribbean settlements.

The earliest European port settlements of the Greater Antilles were oriented externally. They were either fortress cities such as Havana, Santo Domingo, or San Juan, strategically located to guard the passage of treasure fleets carrying plundered New World wealth back to Europe or settlements of buccaneers as at Port Royal, Jamaica, established as an entrepot for the loot acquired by pirates and privateers (Pawson and Buisseret 1975: 20). But the influence of these coastal settlements hardly penetrated the early countrysides. Even in Spanish Hispaniola, jurisdiction emanating from the colonial coastal settlements had little influence over the rural plantation where the sugar mill-owner (*señor de ingenio*) was, for all practical purposes, absolute master over the land tributary to his small sugar-grinding factory (Ratekin 1954: 16).

On the small British and French islands in the eastern Caribbean in the seventeenth and eighteenth centuries, the coastal "urban" areas were most often small complexes of wharfs, warehouses, and stores devoted exclusively to the export of plantation sugar and the import of plantation supplies. To be sure, a merchant class eventually dominated these urban zones and built opulent houses there, but the merchants were rarely considered on the same economic or social level as the large-scale planters of the countryside (Knight 1978: 114). Local economic "development" during the colonial era of the Caribbean was thus accomplished on rural plantations, not derived from local urban centers.

Of course the external market focus of the Caribbean plantation was its lifeline and reason for existence, but it is worth emphasizing that coastal urban sites in the Caribbean grew up to serve the rural zones, not the other way around. In this way, as in many others, the Caribbean differs from later colonial "development" such as in Africa where European port cities were the initial termini of roads and railways penetrating the non-Western interiors, and the cities were thus development nodes (Taaffe, Morrill, and Gould 1963). The contrast between the Caribbean and Africa merits emphasis because the latter "model" is occasionally employed as a textbook case in order to explain the evolution of transport networks throughout the "developing world" (e.g., English 1984: 47).

Sugar cane always has been the overwhelmingly important crop of the Caribbean plantation, although when sugar prices have been low or when the terrain has been unsuitable, other plantation crops have appeared on most islands. The earliest English planters of Barbados and Jamaica, for example, experimented with tobacco and indigo before sugar cane took over there (Dunn 1972: 53–54, 168–71). Coffee had become exceptionally important on upland plantations of St. Domingue (Haiti) in the decades before revolution (Girault 1985: 70–71). Cacao was a forest plantation crop in Trinidad before the British took the island from Spain, and it still is cultivated there (Wood 1968: 97–104). Pimento and ginger offered slight competition to sugar cane in Jamaica before slave emancipation (Higman 1976: 214). Cattle ranching was important on some large Cuban plantations before the intensification of sugar in the 1800s (Knight 1970: 28). And cotton has been cultivated from time to time as a plantation crop throughout the Caribbean region.

Despite the pressure of competing plantation crops in various places and at various times, an understanding of the physical layout of the Caribbean plantation necessarily involves a discussion of some of sugar cane's agricultural and economic characteristics. And an understanding of the layout of the colonial Caribbean plantation, along with the infrastructural elements of roads, buildings, and factories that supported it, goes far in explaining the enduring settlement imprint that still influences most places in the region.

Sugar cane (Saccharum officinarum) technically is a grass and is propagated from plant cuttings (although several subsequent crops or "ratoons" may be obtained from a single planting). Under proper conditions of heat, moisture, and soil fertility the plant grows vigorously for the first month or six weeks, during which time it develops a complex root system and produces leaves (the so-called "tillering phase"), thereby anchoring the cane and shading its undergrowth. Sugar cane matures at anywhere from nine to eighteen months. The stalks of mature cane, an inch or two inches in diameter, then are chopped and squeezed or ground in order to extract the juice. Once cane is cut, the juice must be extracted within 24–48 hours or else its sucrose content begins to fall. Hence, the proximity of a sugar mill (where juice is extracted) is vital to any commercial sugar-cane producer.

After the sugar juice is extracted it is subjected to heating or "boiling" in order to evaporate the excess liquid and to create crystals of the "raw sugar" which are of a granulated consistency and yellowish-brown. At this point the product has, for all practical purposes, assumed its final weight. But subsequent chemical refining steps are necessary to purify the product and to give sugar its characteristic granular, white appearance whose image comes to mind when any European or North American thinks of "sugar." The final refining steps, converting the brownish raw sugar from Caribbean plantations into the value-added, marketable sugar with which residents of metropolitan countries are familiar, have, not surprisingly, nearly always been accomplished in Europe or North America. The metropolis, not the Caribbean, has thereby been the source of marketed sugar, a fact of momentous economic significance.

Milling and initial boiling of the sugar juice was, until the nineteenth century, usually accomplished on a single Caribbean plantation. And the weight and bulk of the exported raw sugar – usually shipped in wooden barrels – meant that external accessibility was a key to any successful sugar-plantation site. The earliest cane cultivation in Hispaniola was in the island's interior, and it was 1517 before the siting of plantations on the south coast "solved the earlier problem of access to markets" (Ratekin 1954:8). One century later, the first sugar-cane plantations of Barbados were concentrated along the island's leeward coast in order to facilitate the loading of the heavy wooden "hogsheads" directly onto lighters to Bridgetown for subsequent shipment to Europe (Watts 1987: 185). Small rowboats carried sugar casks to Georgetown in the Demerara River estuary from the estate factories along the Guiana coast in the early nineteenth century, factories that were located at the littoral. Then the sugar was loaded onto European-bound vessels (Bolingbroke 1807: 31).

Plantations were not always limited to coastal locations, although the development of sugar-cane areas in the interior of larger islands usually had to await the construction of all-weather roads. In the 1740s the Jamaican Assembly began to allocate funds for the extension and improvement of plantation roads in the interior (Craton and Walvin 1970: 58–59). Travel in Trinidad's backlands was exceptionally difficult until some of the trails were covered with local asphalt in the middle of the nineteenth century (Wood 1968: 23–24). The first railway in South America was laid down in 1848 along British Guiana's coastal plain, in part to facilitate the transportation of raw sugar.

The need for cane quickly to be processed precluded sprawling sugar-cane estates, at least in the early days. The typical early Caribbean sugar-cane plantation was a compact unit and not overly large (perhaps 200–300 acres.) The land normally was selected for its flat or rolling character where cane grew best and accessibility was easiest although, when prices were high, sugar cane was sometimes grown on hillsides. Usually the flat cane fields had a gridded appearance because they were intersected by cart trails and firebreaks. The geometric layout of the early cane plantation was, of course, consistent with its quasi-industrial character and thereby aesthetically more pleasing to its European owners than was either the entangled disorder of either aboriginal slash-and-burn plots or natural vegetation (Dirks 1987: 16–17).

Unlike the hacienda of New Spain, where land use often was extensive, geared to small local markets, and reflected owner prestige as much as monetary profit, the Caribbean plantation was cultivated intensively (Wolf and Mintz 1957). In general, land suitable for cane cultivation was planted in cane on the Caribbean plantation, a seemingly obvious point that was especially true when European sugar prices were high. Food for the slaves, after all, could be imported or grown on hillsides, and if a precarious subsistence base for the slave labor force led to their malnutrition and early death, as it often did, more slaves always could be purchased. Yet callousness taken to the extreme could be economically counter-productive, and Caribbean planters realized that the successful output of raw sugar from an estate unit often involved some diversification in land use. So the earliest Caribbean sugar-cane plantations, despite intensive cultivation, devoted surprisingly little acreage to sugar cane itself. On early sixteenth-century Hispaniola, where "a good plantation might hold outright two hundred acres," half of that total usually was left in forest in order that lumber and fuel might be obtained within the plantation grounds. After taking kitchen gardens, buildings, and even saw mills into account, the early two-hundred-acre plantation of Hispaniola sometimes devoted as little as thirty acres to the cane itself (Ratekin 1954: 14).

The percentage of plantation land devoted to sugar cane was much higher in Barbados two centuries later. On Newton plantation in Barbados in 1796, fully half of the estate's 458 acres were in cane, the remainder used for buildings, provision plots, corn, and pasture (Handler and Lange 1978: 65–66). Plantation canelands on Martinique routinely were turned over to grazing after several ratoon crops or sometimes even left fallow (Kimber 1988: 174). On Guadeloupe perhaps one-third of the area of a typical sugar-cane plantation was planted in cane, with a larger area for wood, pasture land, and slave provision plots (*jardins à nègres*) (Lasserre 1978 I: 355).

Examples of cane acreage percentages from various plantations in different islands at different times do not reveal fully, however, the intensity of the cultivation itself. The plantation units usually were large enough that marginal patches of land – gullies, slopes, poorly drained zones – increased the estates' formal total acreage, thereby lowering canelands as a percentage of the total. That plantation cultivation was intense is testified to in the case of the so-called cane "holing" technique that began, apparently on Barbados, early in the eighteenth century in response to deteriorating soil fertility. Gangs of slaves literally rebuilt the topsoil each season by digging shallow square holes, two to three feet on a side, six inches deep, spaced two to three feet from one another, and depositing cane trash and animal dung in the holes. Each slave was expected to dig 60–100 of these holes daily. This backbreaking technique of "soil conservation" was adopted by some of the other planters throughout the British Caribbean, including Jamaica and the Leewards, by 1739 (Watts 1987: 402–5).

The buildings of the plantation nucleus included the grinding mill, boiling house, and the owner's or overseer's dwelling. Larger estates had a variety of other structures. In 1794 Worthy Park, Jamaica, for example, had both a plantation Great House and an overseer's house plus separate buildings for offices, a slave hospital, cattle pens, a curing house, and a still house, all located within one quarter mile from one another at the heart of the plantation (Craton and Walvin 1970: 108–9). The quality of the planters' houses varied throughout the Caribbean region and from one estate to another, from modest wooden dwellings to opulent manor houses and, although the term "Great House" often was used to describe the principal residence of a Caribbean plantation, it would be misleading to portray these dwellings as uniformly magnificent structures (Dunn 1972: 287–96).

Little is known of the plantation slave quarters themselves, besides the certain fact that they were hovels compared with dwellings for most whites. On Barbados in the late 1700s the slaves inhabited wattle-and-daub huts with packed dirt floors and cane-trash roofs. A century later, some of the Barbados slave huts had stone walls and possibly wood shingles (Handler and Lange 1978: 51–53). On Martinique, clusters of slave cabins usually were located downwind from the planter's dwelling and next to the animal pens so that the slaves could watch over the animals at night (Tomich 1976: 104). Slave cottages throughout the Caribbean were situated sufficiently close to the other plantation buildings so as to facilitate slave

control, but often the cottages were relegated to marginal environmental zones. On St. Kitts, slave dwellings were alongside the dank
ravines or "ghauts" that intersected the canelands (Richardson
1983: 64–65). On the coastline of British Guiana, the plantation
owner's house and factory buildings usually were elevated slightly
above the tidal flats on sandy ridges whereas slave houses were
closer to the seawalls and therefore at greater risk in case of possible
ocean flooding.

The production cycle within the boundaries of a given plantation
unit followed a predictable routine of monotonous labor heightened
during planting and then again at harvest time. "Crop time" or
harvest on a typical Caribbean sugar-cane plantation began, as it
does now, at the beginning of the calendar year, after the cane had
been growing for twelve months, and harvest then lasted until May,
June, or even July. The efforts of the field gangs were then coordinated with those in the factory so that the cane could be cut and
transported as rapidly as possible yet not create a damaging and
wasteful backlog at the mill. Auxiliary activities – loading cane trash
into storage sheds, coordinating a supply of fuel, keeping animals fit
and well-fed, maintaining a clean mill yard – all required discipline,
planning, and a keen sense of timing, the organizational hallmark of
industrialization.

It was obvious on even the earliest Caribbean sugar estates that
the level of efficiency in converting sugar cane to raw sugar depended
upon the local application of the most modern techniques available.
The Spanish on Hispaniola constructed a modern sugar mill in 1517
along the Nigua River, west of the capital town. The crushing
apparatus itself was three upright rollers powered either by an
animal (the *trapiche*) or running water (the *ingenio*), the latter the
preferable power source because water-powered mills could produce
an average of 125 tons of sugar each year (Ratekin 1954: 7). The
three rollers allowed double-crushing of (and therefore more juice
from) individual canes, a technical improvement over earlier Mediterranean grinding machines (Galloway 1989).

The three-roller mill constituted the basic internal grinding mech-
anism for nearly all Caribbean sugar factories into the early nineteenth century, although it underwent modification and
improvement as in Jamaica in 1794 where the "lantern wheel"
increased the hourly output of cane juice and narrowed the gap
between the local efficiency of animal- and water-powered factories
(Barrett 1965: 155). Windmills were introduced into Barbados to
power sugar factories probably in 1647 (Watts 1987: 411), almost

as soon as cane first came to the island, and by late in the same century all the biggest sugar planters of Barbados were grinding with windmills (Dunn 1972: 92, 94).

The particular inanimate power source for individual Caribbean sugar mills depended primarily on local environments. Windpower was adopted most readily in Barbados, Antigua, and on the windward slopes of the British Leewards, places where winds were reliable and not subject to retardation by mountains or excessive forest growth. The French were better known for using water, especially on St. Domingue where the same stream often was used to power cane-grinding mills and also as a source of irrigation (Watts 1987: 409–10). But insular particularities overrode cultural differences. The French had windmills on the eastern half of Guadeloupe, and water-powered mills were the most common source of power for sugar mills on British (but, more importantly, mountainous and rainy) St. Vincent into the last years of the nineteenth century. Where neither water nor wind was reliable, horses and oxen powered Caribbean cane-grinding mills well into the nineteenth century.

On January 11, 1797, a steam engine ground stalks of sugar cane at the Seybabo mill in Cuba (Moreno Fraginals 1976: 36), and steam-powered mills arrived in the islands of the British and French Caribbean in the same decade (Watts 1987: 421). The initial trials for these machines all were somewhat unsuccessful, but forward-looking entrepreneurs probably realized that, once modified, the steam engine provided with sufficient fuel could crush canes endlessly, regardless of wind or terrain. Ever-improving steam-powered sugar-cane milling made steady inroads against the older methods throughout the region during the nineteenth century. Steam mills were more likely to be erected in the "newer" cane-growing regions such as in the Guianas and the Greater Antilles, whereas in some of the older and smaller islands, wind power continued as the dominant means of cane milling well into the twentieth century.

After crushing, the second half of the sugar-production process was the "boiling" operation. Extracting sugar crystals from sugar juice involved two simultaneous changes during heating: "clarification" or the removal of impurities; and subsequent evaporation of excess liquids. Clarifying or "striking" the bubbling sugar liquid was a crucial event on every Caribbean sugar plantation, and historical accounts abound with descriptions of how important experienced slave boilermen were in knowing exactly when to cool or to add alkali to the frothing mixture. One eighteenth-century slaveholder of St. Kitts warned fellow planters to treat the boilermen with utmost care as they were "more perfect in their business than any white men can pretend to be" (Pares 1950: 118).

From about 1650 to 1850 on most Caribbean sugar plantations, the boiling techniques changed very little (Barrett 1965: 159). Factory laborers heated the juice in a series of pans or containers in order to create sugar crystals. Then boilermen drained away molasses, an inevitable by-product whose volume varied as to differing boiling techniques. Planters fed the molasses to slaves and livestock, sold it for rum manufacturing, and sometimes discarded it. The invention of the vacuum pan early in the nineteenth century changed things; more sugar juice could be evaporated at lower temperatures without scorching, and the volume of the molasses by-product was reduced accordingly. The first use of the vacuum pan on any Caribbean sugar cane plantation was at Vreed-en-Hoop across the Demerara River from Georgetown, British Guiana, in 1833 (Deerr 1950: 467).

The shipment of high-quality sugar crystals to England derived from the application of this new vacuum-pan technology was not, however, uniformly applauded in the "mother country"; much more important, it was pronounced subject to punitive import duty so as to reduce competition faced by Britain's domestic sugar-refining industry. This event typified colonial control over the Caribbean plantation. Not only was the colonial Caribbean plantation established to produce tropical staples for the metropolitan market, but it also was a truncated enterprise, one whose internal dynamics were controlled, modified, and occasionally encouraged by a constellation of changing rules and restrictions imposed from afar. That control, perhaps it is needless to say, affected directly Caribbean plantation technology: in 1852 Governor Sir Henry Barkly of British Guiana exclaimed that, without discriminatory British sugar duties imposed on semi-refined Guianese sugar, every estate in the colony would have employed vacuum pans like those at Vreed-en-Hoop (Deerr 1949: 468).

Officials in the metropoles rarely tired of condemning Caribbean colonial planters as greedy, rum-soaked opportunists whose production methods were inefficient, old-fashioned, and characterized by "wastefulness and slovenliness" (e.g., Beachey 1957: 61). But not all of the evidence from the Caribbean colonial era supports this view. Consider the remarkably early application of the steam engine for Caribbean cane-grinding, barely two decades after James Watt had patented the machine. When Caribbean planters were uninhibited by metropolitan restraints, they were as innovative and up-to-date as anyone. As another example, Jamaican planter George Price installed a light railway to haul cut canes on his estate in 1844 (Craton and Walvin 1970: 219–20). Washington and New York, it may be recalled, were not linked by a rail network until 1838, a connection still interrupted by ferry boats and intracity stage coaches at the time.

The adoption of modern plantation techniques in the colonial Caribbean was therefore controlled by metropolitan sanctions and mandates, an issue discussed in the following pages. Elements of modernity appeared in the Caribbean at surprisingly early dates, ample evidence that local planters were by no means loathe to accept modern innovations. But these were only elements, not development itself, the latter jealously confined to metropolitan areas. The presence or absence of steam engines or railroads, contemporary geography textbook accounts of the "diffusion" of the industrial revolution notwithstanding, were not surrogate measures of development. The nature and direction of economic growth in the colonial Caribbean and the movement and distribution of its material manifestations, furthermore, were not processes propelled across geographic space by an unseen economic hand, diffusing ever outward like ripples in a pond. They were events controlled by men (Blaut 1987).

### Trade, war, and politics

The loaves or cakes of raw sugar produced in Hispaniola in the early 1500s were either consumed locally or exported to Europe for final processing (Watts 1987: 113–14). In the 1400s the center of European sugar refining had gradually shifted from Venice to Antwerp, but thereafter Amsterdam had become Europe's principal sugar-refining center. By 1661 Amsterdam had sixty sugar refineries (Deerr 1950 II: 453). In that century the Dutch refined sugar emanating in part from their own New World territories: they did not relinquish all of north-eastern Brazil to the Portuguese until 1654, and they took Suriname, and its promising sugar-cane industry, from the English in 1667. But much of the raw sugar hauled by Dutch cargo ships to Amsterdam came from the English and French Caribbean islands. One of the reasons sugar cane had taken hold so rapidly in English Barbados in the 1640s was that the Dutch had introduced the cane, instructed locals in planting techniques, made slaves available, and then transported the semi-refined sugar to the Netherlands for final processing (Dunn 1972: 61–62).

In the ensuing decades, much of the history of the Caribbean region, indeed much of the history of the European nations that competed for domination of the Caribbean, focused on the competition for the refining and marketing of Caribbean sugar. The final processing, as noted above, was almost always reserved for the European states themselves. And the marketing of sugar and also the reexport trade for sugar led to unheard of material wealth for European individuals, companies, and nations. European food-consumption patterns were transformed by Caribbean sugar, and even the whiteness of refined sugar was perhaps "a symbolically potent aspect of sugar's early European history" (Mintz 1985: 22). Yet the Caribbean plantation colonies, in contrast to Europe's expanding economies, remained ossified production zones for raw sugar, a primary producer role eventually undermined by new production techniques elsewhere that would reduce many of the Caribbean islands to inauspicious colonial backwaters by the late 1800s.

It took neither France nor England long to impose legal sanctions against Dutch shipping in the seventeenth century. Dutch traders also had facilitated the introduction of cane into the French islands, as they had to English Barbados, and in the 1650s Dutch navigators were shipping sugar from both Guadeloupe and Martinique to Amsterdam and controlling this traffic to the point that French officials felt their own Caribbean planter colonists were "more the subjects of Holland than of France" (Tomich 1976: 28–29). French minister Jean Baptiste Colbert intervened, mandating that trade between French colonies and the mother country underpinned the economic well-being for residents of both places. Accordingly, in 1664 he declared the trade between the French West Indies and France a monopoly of the French West India Company created for that purpose (Mims 1912: 68).

Colbert's mercantile policy was aimed directly at breaking the Dutch shipping and sugar-refining monopoly, and he initially encouraged sugar refining in the French islands themselves. Five refineries existed in Martinique and Guadeloupe by 1680, benefiting local planters, providing work for white French artisans, and reducing trans-Atlantic shipping costs (Watts 1987: 261–62). But Colbert's encouragement of Caribbean sugar refining soon turned to discouragement, marked by punitive import duties, because of the interests of the incipient (European) French sugar-refining industry.

Centered in the northwest of France, especially at Rouen, the production of granulated (Caribbean) sugar stimulated a vital export trade so that by 1700 France's single most important export commodity was refined sugar (Deerr 1950: 457).

The English took similar measures to thwart Dutch trade, enhance their own and, as a long-run corollary, inhibit real development in their Caribbean islands. The famous Navigation Acts beginning in mid-seventeenth century restricted the colonial trade of London's Caribbean colonies, reducing the commerce between the latter and foreign areas and also inhibiting colonial manufacturing (Sheridan 1973: 41–44). English sugar refining, mainly at factories in London (Deerr 1950: 458), then boomed in a manner similar to its parallel growth in France because of similar restrictive mercantile law. English government officials and parliamentarians acknowledged colonial protest against the trade restrictions, but to allow refining in the colonies would reduce English commerce and shipping. As to the refining industry itself: "(W)e do gain also here in England the great trade of sugar-baking, by which a very great stock and number of people are employed here, a great consumption of coals, victuals, and other necessaries for carrying on of this manufacture" (Williams 1970: 165).

Prohibitions against colonial sugar refining or, more accurately, mandates that the product was only to be semi-refined in the colonies, left varying quantities of molasses to be disposed of on the plantations. Molasses' only real commercial importance was as the raw material for rum. Earliest Caribbean rum was consumed locally. But molasses and its distillate thereafter became crucial trade items in themselves, involving New England where rum was manufactured from molasses, lubricating the West African slave trade and becoming standard fare for metropolitan soldiers and sailors, in times of peace but especially in times of war. The eighteenth-century North American appetite for West Indian rum, both as an item to consume and to reexport, led to the imposition of British legal constraints similar to the Navigation Acts. In particular, New England traders preferred a molasses trade with the French, rather than the British Caribbean colonies; French molasses was cheaper owing to rum's competition at home with French wines and brandies, and North Americans could reciprocate with a surfeit of lumber and corn for planters on French islands. In any case, the British Parliament attempted to control this trade with the Molasses Act of 1733, a law subsequently enforced with varying degrees of success (Sheridan 1973: 339–59).

In response to the seventeenth century French and English mandates and laws against Dutch shipping, the Dutch sea trade in the
Caribbean was reoriented rather than eliminated. The eighteenth-
century Caribbean became known in the Netherlands as the era of
the *kleine vaart* (short journey) whereby Dutch ships compensated
for sanction-induced reduction of the trans-Atlantic *grote vaart* by
transshipping salt, tobacco, cacao, rum, hides, lumber, corn, raw
sugar, and innumerable other items throughout the Caribbean area.
Dutch sea-traders included North America and especially the Spanish Mainland in their itineraries, virtually ignoring the many sanctions imposed against them and thereby playing a major role in
supplying the growth of the eighteenth-century plantation regime
throughout all of the eastern Caribbean. Although Dutch vessels
ranged throughout the region, the center of the *kleine vaart* trade
network was located at Dutch Curaçao and then later at Dutch St.
Eustatius. In 1744 no fewer than 1,240 vessels called at the latter
Dutch island, including 249 from the French islands and 558 from
those controlled by Britain (Goslinga 1985: 204–5). Sugar-cane
planters from the Netherlands, furthermore, had by no means left
the regional cultivation of sugar cane entirely to the French and
English. By 1737 Dutch sugar-plantation settlements worked by
African slaves extended far upriver in Suriname. The Dutch by that
time also had begun to establish similar plantation colonies in the
continental coastal lands to the west, all the way to the Essequibo
River near Spanish Venezuela (Deerr 1949: 208–12).

The importance of Dutch and North American trading notwithstanding, the character of each of the insular plantation colonies was
imprinted by seventeenth- and eighteenth-century trade restrictions.
In general, local manufacturing – not simply of sugar but also of
items as simple as nails and barrel hoops – was prohibited by a
complex system of duties and fines. The French government insisted
that colonies were commercial extensions of the mother country and
should be, literally, fed by the flour and wine sent from Bordeaux
rather than by cheaper North American fare (Watts 1987: 262). The
British imposed similar edicts and constraints including laws controlling the nature of colonial port settlements, fearing that a growth
of urban places would inspire local manufacturing to the detriment
of British-based industries. Inevitably the relentless severity of European trade restrictions led to complaints from local planters. And
the Dutch *kleine vaart* trade plus the similarly illicit activities of
North American traders helped to circumvent a trade involving
duty-laden, high-priced goods (Fortune 1984). Not all Caribbean

colonies experienced similar trade restrictions. In 1716, Denmark began to allow residents of St. Thomas a freer trade than that stipulated by the monopoly held by the Danish West India Company, acknowledging that Denmark was unable to provide the same range of goods as Britain and France and therefore could not insist on trade restrictions as prohibitive as those imposed by London and Paris (Williams 1970: 173–74).

By the early 1700s all of the plantation islands of the eastern Caribbean had been irrevocably caught up in a European-focused sugar trade. And it probably occurred to planters and local officials only rarely that these islands – on which the Caribs had thrived less than a century earlier – could be used for anything other than growing cane. Indeed, the deforestation of all but the highest elevations now necessitated trade to obtain firewood and construction lumber. The high prices for sugar (Watts 1987: 269), furthermore, influenced the intensity and extent of sugar-cane acreage, in turn requiring the continuous importation of food. Roads, buildings, settlements, and local labor forces all supported cane cultivation and were in turn linked closely with a complex, external trade network. The idea that they resided on "sugar islands" perhaps even extended to the slaves themselves; an objective of an aborted slave rebellion in Barbados in 1692 was to take over, not simply destroy, white-owned sugar-cane plantations with a possible intent toward future black-controlled sugar production on the island (Craton 1982: 114).

Caribbean sugar-cane plantations were also helping to create important changes in Europe as well. By the early eighteenth century the average Briton was consuming more sugar than ever before because of the growing popularity of tea and coffee (Sheridan 1973: 27–29). And through the remainder of the eighteenth century and especially into the nineteenth, these beverages – sweetened with Caribbean cane sugar – provided quick energy and food substitutes for Europe's burgeoning proletariat. "Tea time" and "coffee breaks," moreover, helped to condition European working peoples to the punctuality and routine eventually demanded by the industrial revolution. In this way, the colonial Caribbean sugar-cane plantation, beyond providing capital surpluses, provided a critical stimulant that helped to feed and even to socialize working-class Europeans in fulfilling their roles in a new economic order (Mintz 1985; Wolf 1982: 332–46).

Rivalries between white Caribbean planters and European-based colonial officials were as inevitable as they were complex. Especially on the French islands, plantation owners resented metropolitan control over colonial commerce (Tomich 1976: 37–39). English Barbadian planters declared the Navigation Acts prejudicial to Caribbean commercial interests almost as soon as they were enacted, the start of a decades-long British colonial struggle that essentially "became one between metropolitan merchants and Caribbean planters" (Williams 1970: 180–81). Yet British Caribbean planters profited mightily themselves from restrictive trade practices, and they protected these profits by extending restrictive trade practices through their own political power in the Caribbean and eventually in Great Britain as well. In return for accepting prohibitions against local manufacturing, English planters, beginning about 1700, benefited from taxation policies in London that virtually excluded foreign sugar from the London market. British Caribbean assemblies, dominated by planters, then pushed through taxation policies curbing inter-island trade (lest French sugar find its way into British markets). Most important, British planters and their commission agents in London soon began to develop the powerful "West India Interest" eventually including members of the British Parliament and officers of the City of London. The West India Interest maintained a watchful eye for decades over such issues as duty rates for imported raw sugar and the prevention of North Americans from purchasing French molasses (Sheridan 1973: 53,58–71). Despite its name, the cause of the West India Interest, it should be emphasized, was not the accrual of benefits ultimately residing in the islands themselves but to individuals and families who were British, first and foremost, who had derived their financial fortunes through the labor of African slaves on Caribbean soil.

Commercial rivalries among European nation states manifested themselves not simply in competing trade laws but also, not surprisingly, in an almost endless state of Caribbean warfare during the eighteenth century. As Spanish power waned in the region, a series of essentially Anglo-French wars began to create an ever-changing geopolitical map of the colonial Caribbean that would not be stabilized until early in the 1800s. Most islands changed hands at least once, and some many times (Table 2). Battles were not fought continuously: more often naval vessels routinely blockaded ports and sea lanes. The blockades were designed to enforce trade laws because military action in the colonial Caribbean essentially was "a branch of business, not only for the colonists who claimed the protection of the navy, but for the strategists who planned the operations and most of all for the sailors who carried them out" (Pares 1936: viii).

Table 2.  *Colonial control in the Lesser Antilles*
(Arrayed from north to south)

|  |  |
|---|---|
| St. Eustatius: | 1636–1672, Dutch; 1672–1682, British; 1682–1690, Dutch; 1690–1696, British; 1696–1781, Dutch; 1781–1784, French; 1784–1810 Dutch; 1810–1816, British; 1816–present, Dutch |
| Antigua: | 1635–1666, British; 1666–1667, French; 1667-independence, British |
| St Kitts: | 1623, claimed by British; most of 17th century, shared between British and French; 1702- independence, British |
| Nevis: | 1628–1706, British; 1706, French; 1706–1782, British; 1782–1784, French; 1784-independence, British |
| Montserrat: | 1632–1667, British; 1667, French; 1667–1782, British; 1782–1784, French; 1784-present, British |
| Guadeloupe: | 1635–1759, French; 1759–1763, British; 1763–1810, French; 1810–1814, British 1814–1815, French; 1815–1816, British; 1816–present, French |
| Dominica: | 1632–1761, French; 1761–1778, British; 1778–1783, French; 1783-independence, British |
| Martinique: | 1625–1762, French; 1762–1763, British; 1763–1794, French; 1794–1802, British; 1802–1809, French; 1809–1814, British; 1814–present, French |
| St Lucia: | 1803-independence, British held by France nine times, the British six times, and declared neutral twice until 1803; |
| St Vincent: | 1627–1762, claimed by British but occupied by Caribs; 1762–1779, British; 1779–1783, French; 1783-independence, British |
| Grenada: | 1650–1762, French; 1762–1779, British; 1779–1783, French; 1783-independence, British |
| Barbados: | 1627-independence, British |
| Tobago: | 1658–1677, Dutch; 1667–1763, variously held by British and French; 1763–1781, British; 1781–1793, French; 1793–1802, British; 1802–1803, French; 1803-independence, British |
| Trinidad: | 1532–1797, Spanish; 1797-independence, British |

*Source:* Henige (1970)

The British established two principal naval bases, at English Harbor in Antigua and at Port Royal in Jamaica. The French, generally speaking, sent out refitted fleets every spring from the continent. By 1783, when the American War of Independence had come to an end, the British, in addition to maintaining their hold on Barbados and the Leewards, had added to their domain the former French islands of St. Vincent, Grenada, and Dominica. The French, on the other hand, had taken control of St. Lucia and also Tobago, the latter island from the Dutch; these latter two islands would however become British by early in the 1800s. Also at that time London extended her domain to include Trinidad, ceded by Spain, and British Guiana, formerly Dutch. And by the late 1700s the British already had won logwood concessions from Spain to establish full-time timber camps on the eastern rim of Central America in what would by 1862 become the colony of British Honduras.

Special mention must be made of St. Domingue (Haiti) during these decades of intense competition for the Caribbean sugar trade. In the latter part of the eighteenth century, St. Domingue was the richest of all Caribbean colonies, accounting for perhaps one-third of France's entire external trade. In the final decade of the century, amidst the turmoil inspired by the continental French Revolution, black slaves ousted the French, repelled a British counter-invasion, and established the black republic of Haiti early in 1804. British slaves also acted as regular British soldiers in Caribbean military campaigns dating from 1795 (Buckley 1979). The acknowledgment of the crucial military roles of Caribbean slaves and black freedmen in this era of warfare is an important reminder that the colonial wars of the Caribbean were not simply romantic naval battles fought by European sailors against an insular backdrop featuring submissive, faceless slaves. Caribbean slaves also were involved in the fighting and thereby helping to shape their own futures, issues discussed at length in Chapter 7.

The establishment, sustenance, and defense of Caribbean sugar-plantation colonies thousands of miles across the Atlantic necessitated, quite obviously, some means of political control by European nations. It should come as no surprise, furthermore, that the different nations exerted this control in different ways. Speaking very generally, the French were more autocratic and ruled more directly from France than did the British. This general difference, of course, mirrored the political differences in the two metropolitan countries, but perhaps it was also because the British had established a larger territorial stake in the Caribbean and felt the need to establish local long-term political systems in the islands rather than simply decree ephemeral rules designed to funnel export staples to Europe (Wallerstein 1980: 101–2).

The resentment toward metropolitan rule felt by French planters was rekindled daily in confrontations with merchants and trade brokers (*commissionaires*) who profited directly from the protected commerce and therefore aligned themselves with France and against free trade (Tomich 1976: 37–39, 158–59). The low-water mark in colonial Caribbean governance systems was perhaps achieved in St. Domingue where local government officials are said to have been animated by little more than a "ferocious cupidity and greed" with attendant waste, bribery, nepotism, and all-round incompetence as the hallmarks of the French colonial bureaucracy there (Lewis 1983: 124–29). And it is perhaps gratuitous to suggest that such an atmosphere, which reinforced important caste distinctions even among white planters, as well as among whites, browns, and blacks, was heartily conducive to the rebellion that subsequently evicted the French from Haiti. After 1848, local political representation was expanded for the remaining French colonies, an arrangement that actually had been experimented with a half-century earlier (Knight 1978: 161).

The legislative assemblies which were, essentially, local planter forums of the early English Caribbean were generally replicated in Grenada, Dominica, and St. Vincent when the latter three were ceded from France to Britain in 1763. St. Lucia, British Guiana, and Trinidad, however, were controlled more directly as "Crown Colonies" when they became part of the British colonial realm in the early nineteenth century. The Crown Colony system meant the near-autocracy of an appointed governor, as opposed to the quasi-autonomous elected assemblies, and Crown Colony rule then became more, not less, widespread in the British Caribbean later in the nineteenth century. The post-emancipation Morant Bay Rebellion in Jamaica in 1865 led to a fear of black control on the island and a subsequent revocation of Jamaica's constitution in favor of Crown Colony rule there. Crown Colony government then engulfed most of the British possessions of the eastern Caribbean, with the notable exception of Barbados, as the British Colonial Office centralized its administrative control over the region (Levy 1980).

Dutch Caribbean possessions came under the control of the Dutch West India Company which was reorganized in 1648, after peace with Spain. The new version of the Dutch company divided the country's overseas possessions into two classes: those the company controlled directly (including Aruba, Bonaire, Curaçao, and the coastal colonies of Essquibo and Demerara); and the others controlled by Dutch cities (Suriname) or proprietorships (St. Eustatius and Saba). The Dutch West India Company itself was controlled by a board elected annually by Dutch government officials (Goslinga 1985). The Danes in 1850 established an assembly of elected members modeled, generally speaking, on the planter legislatures of the older British colonies. Representatives were elected on a proportional basis from St. Croix, St. John, and St. Thomas, by electors who possessed strict age (25) and income ($500/year) qualifications (Williams 1970: 397–98).

The character of political control in each Caribbean colony was a function of its ties to particular European nation-states but, equally important, its changing economic circumstances. The contrast between the political changes in British and Spanish Caribbean colonies in the late nineteenth century exemplifies the point. The shift to Crown Colony governance for nearly all the British islands has been interpreted "as a means of protecting the interests of European planterdom" (Green 1976: 353) in an era of plantation decline and an associated rising local influence of black and brown men. At the same time, local planters and merchants on the Spanish islands were, in the late nineteenth century, realizing a growing freedom from Spain, with an increasing self-governance. These latter changes were related directly to the intensification of sugar-cane production in the Spanish Greater Antilles at the time, intensification that, ironically, was furnishing part of the worldwide increase in cane-sugar production leading to the relative decline of British Caribbean plantation colonies (Mintz 1974: 86–94).

The terms "self-governance," "freedom," "assembly," and "legislature," when applied to the colonial Caribbean, bear little relationship to the kind of participatory democracy these words connote late in the twentieth century. Locals indeed cast votes in the colonial Caribbean, but the right to vote was reserved solely for those who satisfied restrictive property and income qualifications. In almost every case these qualifications ruled out all except white planters, merchants, and government officials until a handful of nonwhite voters began to appear in the nineteenth century. Actually, conditions for the black working classes of the colonial Caribbean probably were worsened during periods of greater local autonomy and "freedom" for local assemblies and legislatures. This is because concern for the welfare of the slaves and, later, emancipated members of the working class generally was felt more strongly in

Europe's socially detached isolation rather than among the colonial Caribbean elite.

The interrelationship between relative decline in the British Caribbean in the nineteenth century and simultaneous surging growth in cane production in the Spanish Caribbean was part of a global acceleration in tropical sugar cane production. The Dutch already had increased this production in Java by the early 1700s, but a century later colonial cane fields were expanding throughout Asia (India, the Philippines), the Indian Ocean (Mauritius), and South America (Brazil). In the Caribbean itself, a dramatic intra-regional shift in sugar production was occurring. In 1815 Cuba produced 40,000 tons of sugar, half the total for Jamaica; in 1894 Cuban tonnage was 1,054,214, four times the total for the entire British Caribbean (Williams 1970:366–67). Much of the Cuban advantage – similar to that for the Dominican Republic and Puerto Rico which were awakening as well – was environmental: the rich, expansive, relatively unused soils of Cuba attracted investment capital much more readily than did the tiny, worn-out British islands. And the Cuban environmental edge, combined with the proximity of the United States market helps to explain the Cuban sugar boom of the nineteenth and early twentieth centuries, an issue discussed further in the following chapter.

In the small islands of the French and British Caribbean that had been the rich commercial prizes of the wars of the eighteenth century, almost the entire nineteenth represented severe economic decline. In the French colonies, especially during the period 1830–1848, decreasing sugar prices and increasing costs led to a prolonged crisis (Tomich 1976: 169). Prices for British West Indian sugar during the entire nineteenth century also declined precipitously (Watts 1987: 495). Boom-and-bust was not unknown from previous years; much of the slave era had seen widely fluctuating sugar prices, even with guaranteed and protected home markets (Sheridan 1973: 389–414). But earlier price changes often had been influenced by weather or warfare, so planters had learned to anticipate that better prices would follow declines when local and external influences brightened as inevitably they would. The same planter attitudes probably prevailed in the first decades of the nineteenth century, especially when brief price boomlets – such as in the British Caribbean in the late 1850s – must have suggested that stability and prosperity soon would return.

Yet the inevitability was not that sugar prices would oscillate but that the era of mercantilism was over and that of industrialism surging forward with an emphasis on mass production and economies of scale. These transformations, in turn, would create massive change in the world sugar market (Albert and Graves 1984). As British industry sought to export its manufactures to a global market, and since so many other tropical areas now were exporting mainly sugar, the protected sugar prices for British West Indian planters became economic anachronisms; these preferential prices were rescinded in 1846. The French attempted to shore up their flagging Antillean sugar-cane industry by establishing central sugar factories. But the eastern Caribbean faced even stiffer competition than that posed by new cane-sugar plantations worldwide. European beet sugar, especially from Germany and France, had begun to take ever-increasing shares of the European market by mid-century. French Caribbean sugar-cane planters bemoaned the competition, asserting that the *pacte colonial*, the special economic relationship between France and her Caribbean colonies, had been broken by favoritism accorded beet producers; continental beet sugar interests responded (as early as 1839) that the ruin of the French beet sugar industry would affect as many people in a single French *arrondissement* at home as there were whites in all the colonial French sugar colonies combined (Tomich 1976: 72, 83).

The small British and French islands – already beleaguered by economies of scale in other tropical colonies – were therefore no longer producing an exclusively tropical staple. Most European nations, furthermore, encouraged improvements in their own local sugar-refining industries with a complex system of refining payments called "bounties." These encouragements stimulated European beet-sugar production, flooded the open European markets with cheap sugar, lowered the prices for Caribbean sugar, and created the "bounty sugar depression" throughout the British Caribbean in the last two decades of the nineteenth century. During 1884, the date usually acknowledged as the depression's first year, the price of Caribbean raw sugar on the London market fell from 19 shillings per 100 lb to 13 (Deerr 1950: 531), a decline caused mostly by a massive dumping of German beet sugar in Britain. An even more precipitous plunge in price (*chute des prix*) occurred in the French islands. The price for a 100 kg "quintal" of sugar exported from Guadeloupe in 1880 was 70 francs, dropping to 25 francs three years later (Lasserre 1978: 403–4).

By the late 1800s, the European sugar market, which the small French and British Caribbean islands had created and sustained for two centuries, really no longer needed their production. It was not that metropolitan demand had decreased. Quite the opposite, the urban working classes in Britain were enjoying an ever-increasing quantity of sugar-supplemented fruit and jam in their diets, items formerly considered luxuries (Hobsbawn 1969: 162–63). But the flood of sugar into London from throughout the tropics and, increasingly, from European producers, had lowered prices. British politicians and industrialists vied with one another in coining slogans to affirm the moral superiority of *laissez faire* economics whereas two centuries earlier similar spokesmen had insisted on trade protection. Now the British jam and confectionery industries, blessed with low sugar prices and accelerating demand for their products, sermonized about a "free breakfast table" for working-class Britons, a God-given right that could not be sacrificed for a protected home market (Deerr 1950: 506–7).

The suffering among the black workers in the old British and French sugar colonies during the depression years was acute, not that they or their ancestors ever had truly prospered from their toil on Caribbean sugar plantations. Local economic alternatives were limited, furthermore, because the islands had been "developed" to produce sugar cane since the early 1600s. Living conditions in the British Caribbean had become so depressed by the beginning of the twentieth century that the British Colonial Secretary, Joseph Chamberlain, characterized Britain's Caribbean sugar-plantation colonies as the "Empire's darkest slum." And few observers, amidst the ascendance of modern mass-production, had time to contemplate the crucial role in recent centuries played by the Caribbean slaves in producing food substitutes for a European working class and also enriching beyond imagination the colonial powers: "Slave and proletarian together powered the imperial economic system that kept the one supplied with manacles and the other with sugar and rum; but neither had more than minimal influence over it" (Mintz 1985: 184).

## Slavery and the slave trade

While producing sugar for European workers' consumption, the generations of black slaves on Caribbean sugar plantations actually had been more intimately dependent on external market circumstances – including swings in prices of agricultural staples – than had their masters. Especially on the smaller islands, slaves relied on supplementary food imports from North America; this reliance was heightened if local weather events reduced outputs from the slaves' own subsistence plots that invariably were relegated to marginal environmental areas in the islands themselves. The autumn hurricane season, when shipping was curtailed in the Caribbean, always was the "hungry time" among the slaves. Then the Christmas season usually provided welcome relief as slave diets were enriched with fresh food imports (Dirks 1975). But if external events prolonged shipping curtailments or raised prices of imported goods prohibitively (whether or not prices were "prohibitive," was, of course, a planter's decision) slaves occasionally starved. The worst known case was during the American War of Independence when naval blockades reduced North American food shipments to the Caribbean. An estimated 15,000 slaves died of famine in Jamaica alone between 1780 and 1787 as did hundreds more in the smaller islands (Williams 1970: 226–27). Involved in a trans-Atlantic enterprise, Caribbean planters were wary of market changes halting the timely flow of goods and profits; their slaves, on the other hand, had even less control over these external circumstances on which they depended, not for profits, but for their very lives.

The handful of African slaves brought to the gold diggings on Hispaniola in the first decade of the sixteenth century were the forerunners of a river of humanity flowing from West Africa to the Caribbean over the next four centuries. Millions of black men and women were torn from their African homelands to toil on Caribbean plantations, thereby creating Caribbean demographic and cultural patterns "in many ways more African than European" (Meinig 1986: 171). Why was slavery the means of Caribbean labor coercion? And why Africans? The obliteration of Arawaks and Caribs created a labor vacuum, and the tradewinds flow from east to west across the tropical Atlantic. So Europeans could somewhat easily remove slaves from a West African breeding ground for labor, the latter region thereby absorbing the cost of producing manpower (Wallerstein 1974: 89). And, although historically oriented scholars are by no means agreed as to why slavery – as opposed to less severe means of coercion – was instituted, a key seems to be that the availability of unoccupied land in a Caribbean frontier region called for the strictest kind of labor control, lest imported workers take up land of their own (Mintz 1977: 255–57).

Whatever the reasons, slavery remained the principal means of Caribbean labor control for centuries, the region's last slaves freed in Cuba in 1886. During the nearly four hundred years of Caribbean slavery, an estimated 4.6 million African slaves were brought into the region: 1,665,000 to the British Caribbean; 1,600,000 to the

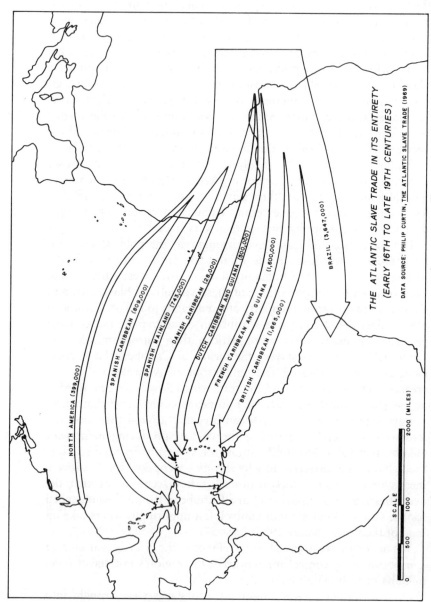

THE ATLANTIC SLAVE TRADE IN ITS ENTIRETY
(EARLY 16TH TO LATE 19TH CENTURIES)

DATA SOURCE: PHILIP CURTIN, THE ATLANTIC SLAVE TRADE (1969)

NORTH AMERICA (399,000)

SPANISH CARIBBEAN (809,000)

SPANISH MAINLAND (743,000)

DANISH CARIBBEAN (28,000)

DUTCH CARIBBEAN AND GUIANA (500,000)

FRENCH CARIBBEAN AND GUIANA (1,600,000)

BRITISH CARIBBEAN (1,665,000)

BRAZIL (3,647,000)

SCALE

0    500    1000    2000 (MILES)

Map 2    The Atlantic slave trade

French Caribbean; 500,000 to the Dutch Caribbean; 28,000 to the Danish Caribbean; and an estimated 809,000 to the Spanish Caribbean. One reason these data astound is that, compared with the estimated total for all African slaves brought to the New World (9,566,000), including those taken to Brazil, the Spanish Mainland, and all of North America, those arriving at Caribbean destinations account for nearly half. These estimates however, mask the uneven temporal flow of the Caribbean slave trade (an astonishing 250,000 African slaves are estimated to have arrived in Jamaica between 1781 and 1810) as well as its complexity: slaves brought to the Caribbean aboard Spanish or Dutch ships in the early years or aboard French or British ships in later decades, for example, often were sold to planters of different nationalities once they arrived in the New World (Curtin 1969).

The oversimplified "Africa to Caribbean" trajectory describing the colonial slave trade is further complicated by the realization that European slaving forts or slave-trading depots extended from Cape Verde in the Senegambia south to Benguela in the Angola region (Ashdown 1979: 15). These coastal regions of Africa, furthermore, were augmented by the Canary Islands and the Cape Verde Islands as origins for slaves, the latter places exceptionally important as a source of some of the earliest slaves brought to the Spanish Caribbean. Different zones along the slave coast varied greatly through time in providing numbers of slaves. And recorded origins such as "Congolese" or "Angolan" hardly distinguished tribal or linguistic identities of individuals coming from the vast catchment areas designated by these and other broad geographical place names.

The horror of the Middle Passage – a journey accented by suicide, malnourishment, disease, and occasional rebellion – did not automatically homogenize captured Africans into docile members of plantation workforces. Caribbean planters recognized slaves' differences and came to associate particular behavioral characteristics with specific African areas of origin. The white planters of Antigua became painfully aware, for instance, of the organization behind the aborted slave revolt of 1736, planning that could be traced to recent Akan arrivals from the Gold Coast (Gaspar 1985: 236–38). But maintaining vestiges of African culture was particularly difficult within the overall system of slavery, especially when planters consciously mixed batches of slaves, as they often did, to avoid the kind of social solidarity that helped fuel the Antigua uprising and others. Once delivered into the hands of a particular planter, furthermore, an African slave or his descendants possibly would be transferred to other islands. Between 1815 and 1825, for example, roughly 20,000 slaves were moved permanently from one British Caribbean colony to another in order to match slave populations with planters' changing labor requirements (Higman 1984: 81).

Perhaps the most troubling revelation derived from the demographic data from the era of Caribbean slavery is that, with few exceptions, large numbers of imported slaves were nearly always necessary to compensate for high Caribbean slave death rates. Europeans originally had brought Africans to Hispaniola, ironically, to compensate for the high death rates among Arawaks and Caribs. But Afro-Caribbean slaves, especially when compared with their counterparts in the United States, were unable fully to reproduce their own populations, thereby necessitating the annual migration of "salt-water Negroes" to replenish plantation labor forces (Curtin 1969: 92–93).

The high slave death rates seem to have characterized all Caribbean slavery with little distinction between islands or colonial powers – data that argue against the idea that slavery within Catholicism was somehow more benign than under English or Dutch Protestant planters (Klein 1967). The (Catholic) French Antilles, for example, received an estimated 135,000 slaves between 1763 and 1776, although the French islands' total slave populations increased by only 85,000 in the same period, deaths thereby outnumbering births by 50,000 (Lasserre 1978 I: 293–94). A salient characteristic of these population losses in the French Caribbean was the exceptionally high rates of infant mortality among the slaves, a regional characteristic that afflicted British and Spanish islands as well. West Africa also suffered terribly high infant and child mortality rates at the time, so perhaps not all blame for high Caribbean slave death rates lies with colonial planters (Kiple 1984: 119).

Yet Caribbean planters decreed work regimens and were responsible for the provision of food – or the means to produce food – for their work forces, and comparative evidence suggests that the actions of plantation owners or their agents could and did exert ultimate control over a given slave's lifespan. The exceedingly high death rates (25 percent per annum) in the earliest years of slavery in the English Caribbean and then again in the "new" ceded islands of the British Caribbean and Trinidad (14 percent per annum) over a century later (Dirks 1987: 91–92), occurred when new lands were being cleared. It follows that the era of ecological devastation described in the previous chapter was notable not only for the irreplaceable loss of much of the region's flora and fauna, but also that it was accomplished by sacrificing a massive amount of expendable human energy imported from West Africa. Compare these high death rates with the notable case of Barbados where, by the early nineteenth century, slaves were achieving a natural increase in population, in part because of the abolition of the British slave trade and the associated necessity for planters to relent in their harshness in order to encourage slave births (Kiple 1984: 106).

Much of the answer as to why Afro-Caribbean death rates were so very high must be because of the region's insular geography. The planter preoccupation with sugar cane, to the detriment of local subsistence agriculture, was only part of it. Probably more important was that environmental events – drought and flood come to mind most readily – created local subsistence shortfalls that could be remedied only by importing foodstuffs. These importations, furthermore, would have had to come from long distances because the same weather events would have affected nearby islands. And when these shipments were curtailed for any one of several possible reasons, slaves starved or, more often, were malnourished. Malnutrition, finally, was possibly reflected in low birth rates, disease susceptibility, and related ills that altogether would have influenced slave populations (Kiple 1984; Sheridan 1985: 200–6). It is obvious that many variables influenced rates of natural population increases and decreases among Caribbean slaves. Their very low birth rates, for example, have been the focus of a good deal of recent scholarship. The most exhaustive statistical work to date dealing with Caribbean slave populations has nevertheless suggested that, in accounting for varying rates of natural population increase in the region, mortality differentials are the key variables (Higman 1984: 374).

The slave subsistence grounds themselves, especially on the larger islands or in places with "idle" lands, seem to have been productive, given reasonable weather conditions. Slaves in Jamaica cultivated bananas and rootcrops and owned their own pigs and cattle. St. Vincent slaves subsisted on vegetables, eggs, and fresh fish from inland streams (Dirks 1978: 134–35). On Martinique slaves are said to have been provided as much land as they could cultivate and were given every Saturday to work the plots (Tomich 1976: 155). And it is not surprising that it is in these tiny hillside subsistence gardens, the only lands over which the slaves had some control, that elements of African culture and lifeways seem to have been most noticeable. Slaves cultivated the ackee fruit, okra, congo bean (blackeyed peas), yam, millet, and sorghum, all of which they had known in West

Africa (West and Augelli 1976: 75). On both British and French islands, slaves distributed surpluses from their subsistence grounds at regularly scheduled markets, thereby earning small amounts of money and providing the historical underpinning for similar markets that still exist throughout the Caribbean region late in the twentieth century (Mintz 1974b: 194–206).

The contrast in work attitudes between the slaves' labor applied to their own subsistence grounds and house gardens versus the drudgery on plantations must have been extreme. Caribbean plantations' slave workforces were divided into various groups or gangs, down to and including small children. Regimentation, combined with strict discipline, marked a routine of repetitive chores that lasted from dawn until dusk and varied little from day to day except during cane harvest. Drivers, who usually were themselves slaves, often kept field workers, literally, in line as they worked through the fields. Planting, clearing, and weeding usually were accomplished by slaves wielding long-handled hoes. Then they used machete-type cutlasses or bills to harvest the crop during the first months of the calendar year (Dirks 1978: 17–21).

At harvest the value of the slave boilermen's expertise became obvious, providing a glimmer of social differentiation among the slaves themselves. Most, not all, slaves were field workers. Slave artisans and fishermen throughout the Caribbean region enjoyed better diets and greater spatial mobility than did plantation field hands (Wilson 1973: 192). It hardly needs emphasis that these privileged slaves were by no means satisfied with their lot because they often were the instigators of slave insurrections.

The balking, malingering, and understandably thorough apathy displayed by Caribbean plantation slaves toward regimented estate labor were interpreted, at least for public consumption, by the European plantocracy as laziness and stupidity. And these pejoratives were almost invariably disseminated in a racist manner because the plantation's transformation of the Caribbean region had been demographic as well as environmental. The Spanish had enslaved Arawaks before the arrival of Africans so Caribbean slavery did not need Africans, any more than modern variations of racism need slavery, to survive. But slavery and imported Africans eventually became so closely interrelated in the region that it was nearly impossible to separate one from the other. And an enduring Caribbean color caste, perhaps best described by the popular regional term "money whitens," may be traced to the region's plantation slavery, although subtleties and nuances in this system of color and

class vary in many ways from one Caribbean locale to
(Hoetnik 1967).

The white versus black polarity on the earliest sugar-cane pla.
tions soon became more complex with the emergence of a mixeo
blood population of brown men and women. This "mulatto" caste
(*pardo* in Spanish, *gens de couleur* in French) had become a
noticeable segment of the human population throughout the region
by the nineteenth century, comprising perhaps 5 percent or even
more of the total (Knight 1978: 97). On the plantations, mixed-
blood slaves often were artisans, house servants or held other
positions elevated from field drudgery. In the years before emancipation, brown men and women were freed as individuals more often
than dark-skinned blacks, and gradation of skin color was identified
during slavery with occupational categories throughout the region,
dark field laborers at one end of the labor spectrum and light-
skinned house slaves at the other. The same rough distinction still
holds in the Caribbean today, a general observation that is by no
means a social axiom.

The presence of mixed-blood peoples provided something of a
social continuum between black and white, although mixed-bloods
did not always fit neatly into the regimented hierarchy of the
plantation regime. So brown peoples resided and worked in urban
areas in disproportionately large numbers. Brown freedmen in
Barbados already had gravitated to Bridgetown in significant numbers in the decades prior to emancipation (Handler 1974: 16–17).
And the mixed-blood peoples of Paramaribo, Suriname, in the 1790s
typically occupied social positions between white and black
extremes; their occupations, furthermore, were probably typical of
their social group throughout the entire region: "They were furniture
makers, carpenters, masons, leather-workers, wood-carvers. They
made primitive pottery and glass-ware, they were repairers" (Goslinga 1985: 522).

The ambivalence of the social status of mixed-blood peoples in
Caribbean slave colonies complicates efforts to determine where
their sympathies lay. To be sure, they often followed white norms of
dress and behavior. Many owned property, including slaves, and
often were faithful members of white-dominated militias whose
principal duty was to quell possible slave disturbances. But brown
people themselves also led the uprisings, their role in the Haitian
rebellion only one example. A further illustration from the British
Windwards was that in 1795 Julien Fédon, a Grenadian planter who
reckoned his descent from both France and West Africa, led a

rebellion that occupied both the local militia and regular British troops for two years (Cox 1984).

A comparatively high percentage of persons of mixed blood inhabited the Spanish-speaking Greater Antilles in the nineteenth century. This was a possible result of decades of quasi-independent development in that area, at the same time that the smaller French and British islands were intensifying sugar-cane production with attendant massive inputs of black African slaves annually. In any case, the laborers of the sugar estates of the large Spanish islands were ethnically diverse – Asian, Yucatecan Indians, black, mixed, white – and their free versus slave relationship to particular estates were similarly varied. In 1872 in Puerto Rico, among the 12,512 slaves on the island, 4,374 were counted as mulatto and 187 white (Nistal-Moret 1985: 147).

## Emancipation and its aftermath

Although the nineteenth-century Caribbean was marked by dramatic economic change, it is most important as the region's century of personal freedom from slavery. The British Emancipation Act of 1833 called for a transition from slavery to apprenticeship in 1834 followed by full emancipation on August 1, 1838. French and Danish slaves were freed in 1848, Dutch slaves in 1863. Emancipation came later in the Spanish islands, in Puerto Rico in 1873 and in Cuba in 1886. As in slave societies elsewhere, the transition from slavery to freedom in the Caribbean was gradual rather than abrupt, and plantation labor coercion continued thereafter (Cooper 1980). Immediately after emancipation, Caribbean planters attempted to control members of the newly freed working classes by restricting their access to local lands, enacting immobilizing vagrancy laws, and importing thousands of laborers into the region in order to drive wages down. The ex-slaves pushed just as hard in the other direction to consolidate their new independence by establishing their own village settlements, emigrating for higher wages, and forming their own informal social networks. It was an historical period in which new geographical patterns were established in an old environment that was clouded by old animosities, patterns that have endured in many ways to the present day.

Events outside the Caribbean probably influenced the timing of emancipation in the region more than did internal developments. The newly emerging tropical markets elsewhere were more promising as destinations for manufactured goods from Europe than were the tiny, fragmented markets in most of the Caribbean where tired soils and old equipment limited profits and therefore buying power. Besides, slaves were nearly penniless and hardly constituted a "market." That was a key argument in favor of emancipation: slavery not only was immoral but also inefficient and expensive compared to the work free men could provide (Williams 1970: 256–58). The economic and political underpinnings of eventual slave emancipation, furthermore, were given moral expression. In 1787 the Anti-Slavery Society was formed in Britain, with William Wilberforce their leader in parliament. A similar society, the *Amis des Noirs* (Friends of Black People), emerged in Paris the following year (Ott 1973: 21). Similar abolitionist sentiment, often expressed by religious leaders and animated by slogans extolling the virtues of free labor, existed throughout Europe and North America. Anti-slavery groups formed late in Spain, in the 1860s, probably in large part because slave-based Cuba was Spain's most valued market (Scott 1985: 284).

Caribbean slaves did not wait passively for their owners to grant them freedom. The trans-Atlantic slave trade from Africa to the British Caribbean and to the United States had been abolished in 1807, and rumors abounded thereafter that freedom was at hand. Slave insurrections raged in British Guiana in 1823 and in Jamaica during the 1831 Christmas season. When British slavery ended in 1834, it was replaced by the four-year apprenticeship period. Only Antigua granted unconditional freedom. On nearby St. Kitts, word spread quickly throughout the slave populace that Britain also had granted them full freedom but that it was being withheld by white St. Kitts planters; the slaves on St. Kitts accordingly withdrew from the plantations into the mountains, only to be evicted forcibly from the highlands by the local militia (Frucht 1975). Preemancipation slave uprisings frightened non-British planters too. On Danish St. Croix, the slaves forced emancipation through rebellion on July 2, 1848. They sacked homes in Fredriksted, chased local whites onto nearby ships, and threatened burning and pillage if they were not set free. The Danish Governor-General capitulated and granted freedom that day (Williams 1970: 326–27).

After emancipation was declared in the Danish islands, planters demanded monetary compensation for their losses in slave property, a precedent established by Britain. The British Emancipation Act of 1833 had called for a £20 million payment to the owners of the 540,000 slaves in British Caribbean colonies, money paid directly to the planters to be used however they wished. Eventually planters on all the Caribbean colonies except Cuba received similar payments. The Dutch system of payment was perhaps the most complex, with varying rates of compensation for sugar cane workers, forest plantation slaves who tended cacao and coffee groves, and domestic slaves (Williams 1970: 332–33).

During the 1834–38 British apprenticeship period, planters feared that, with full freedom, ex-slaves soon would abandon plantation work altogether. Apprenticeship itself, moreover, was not a time of healing between planters and their paid apprentices. Rather, it was marked by claims and counter-claims between planters and workers with new work and pay arrangements adjudicated by a limited number of special magistrates (Burn 1937; Marshall 1977). When unconditional freedom finally was granted to British slaves on August 1, 1838, some freedmen left the plantations although the predicted wholesale flight never really occurred. When freedmen did leave estate grounds, it usually was because of harsh contemporary planter attitudes and behavior rather than freedmen's memories of the brutality of slavery (Hall 1978). In any case, free village settlements emerged throughout the British Caribbean at emancipation, communities in which a newly freed people began to adjust to a new beginning from a past distorted by slavery. Members of this reconstituted peasantry (Mintz 1974b: 146–56) created new farm plots, erected dwellings, and began to offer themselves as part-time workers on nearby sugar estates. And the dual role of part-time (peasant) subsistence cultivator and part-time (proletarian) wage laborer established by Caribbean freedmen at emancipation created a complex social type that social scientists continue to explore one and one-half centuries later (e.g., Frucht 1967; Mintz 1953b; Trouillot 1988).

The thousands of new hillside house plots and mountain farms emerging in the British islands in the decade after slavery had a distinctive environmental dimension, so much so that "highland adaptation" has been coined by Sidney Mintz (1974b: 234–36) in attempting to capture the ecological essence of these new settlements. Along the coastal plain of British Guiana, in contrast, a "lowland adaptation" to freedom saw ex-slaves establish their own independent villages on defunct plantation grounds; these new lowland communities were outside plantation boundaries but close enough that ex-slaves could walk back and forth to the estates (Farley 1954).

Proximity to the wage labor provided by plantations was crucial in both highland and lowland village settings. Sheet erosion and drought often plagued highland communities, thereby creating sub-

sistence shortages that could be supplemented with periodic plantation work. The villages of black freedmen along the Guianese coastal
plain were on defunct plantation lands with canal systems originally
laid out and operated under central control. Drainage and irrigation
problems arose almost immediately in these new coastal villages
(Young 1958: 26), sending freedmen out to work on plantations.
The environmental marginality of the new communities throughout
the British Caribbean thus reinforced an ongoing interdependence
between planters and village workers at emancipation, one needing
cheap labor, the other seeking a means to supplement subsistence
uncertainties.

Although emancipation came later in both the French and Spanish
islands, planters there also predicted a loss of their labor force,
similar to the laments of British planters, when freedom was granted.
Released slaves on Martinique left plantations in large numbers and
began to colonize mountain lands almost immediately after emancipation, and an improved road network on Martinique subsequently
provided access between individual land parcels and ongoing estates
as on the British islands (Kimber 1988: 220–21, 226). Cuban
planters bemoaned the impending loss of their slaves in the 1880s
when emancipation finally came. And, although former slaves in
Cuba sought the best wages they could find and, in so doing, often
left their former masters' estates, "abolition did not trigger a
catastrophic flight of former [Cuban] slaves from plantation labor"
(Scott 1985: 239).

In some islands after emancipation, lands were unavailable for
viable small-scale settlements, even of the part-time variety. A lack
of available village land was especially critical in the few cases
where planters dominated entire islands (Bolland 1981). On tiny St.
John in the Danish Caribbean, the slaves were delighted when
notified of their freedom on July 6, 1848, but they soon found
themselves enmeshed in a web of new restrictions regulating both
their access to local lands and also their personal movements. It is
therefore not surprising that "Some laborers began fleeing . . . when
they realized that there was no other way to escape the plantation
systems" (Olwig 1985:88). Freedmen responded similarly in St.
Kitts and Barbados, both British islands, where planters continued
to occupy or control nearly all of the local lands at emancipation
(Marshall, 1968: 254). From these and other small islands, freedmen faced with imposed landlessness emigrated in small sailing
sloops and schooners in a "migration adaptation" (Richardson
1980) at emancipation. They thereby initiated a migration tradition

which persists today throughout the Caribbean, a tradition discussed in Chapter 6.

A system of sharecropping, locally referred to as *metayer* or
*metayage*, developed in some of the small British islands – especially
in Nevis, St. Lucia, and Tobago – in the decades following slavery,
providing further evidence of the variation with which Caribbean
peoples and insular lands were matched after emancipation. Planters
and freedmen signed contracts specifying that the latter would
cultivate sugar-cane lands for a fraction (usually half) of the output
and the former would provide transportation and access to milling
facilities. *Metayage* was a means by which planters could withstand
low sugar prices while retaining control over the land. It was also an
outlet for complaints and bickering by the planters over alleged
misuse of the soil from poor cultivation practices and, on the other
hand, by freedmen concerning alleged underpayments from planters.
*Metayage* seems to have given black freedmen at least a partial stake
in local resources, at least in Nevis where sharecroppers were
cultivating a large proportion of the land in foodcrops by mid-
nineteenth century (Marshall 1965; Richardson 1983: 99).

The ability of planters to coerce free laborers beyond emancipation and the variety of options available to black freedmen to resist
this coercion depended in large part on local population densities. A
greater access to land increased freedmen's local economic choices
compared with the restricted possibilities in places where land was
limited or unavailable. The only way to adjust local population
densities immediately to favor the plantocracy, obviously, was to
import laborers from elsewhere, precisely the strategy that many
planters followed (Green 1984: 116). Caribbean planters convinced
the home governments of post-emancipation labor "shortages" only
after endless complaining that freedmen were now unreliable and
hard to find. The reality may have been quite different. Black
laborers throughout the Caribbean after slavery were beginning to
organize into labor groups, collectively appealing for higher wages
to anyone who would listen (e.g., Adamson 1972: 34).

The "solution" to perceived labor shortages came first from the
Indian subcontinent as Caribbean planters enlarged local labor pools
with immigrant workers and thereby compounded the region's
cultural complexity. The British Colonial Office approved bringing
indentured workers to the Caribbean from India under five-year
labor contracts starting in 1838. From then until 1917 over 400,000
men and women were shipped from recruiting terminals in Calcutta
and Madras to work on British Caribbean plantations; their desti-

nations were mainly British Guiana (240,000), Trinidad (135,000), and Jamaica (33,000), with small numbers sprinkled throughout the other islands. The French and the Dutch imported workers from India into the Caribbean as well. Between 1852 and 1885, France brought nearly 100,000 Indians to Guadeloupe, Martinique, and French Guiana, some from the French Indian enclave at Pondicherry and many aboard British ships. Roughly 35,000 Indians came to Dutch Suriname between 1872 and 1917, and another 22,000 Javanese arrived in Suriname in the first two decades of the twentieth century (Tinker 1974). Black freedmen saw the newcomers as wage competitors. The antipathy among these competing, ethnically different laboring groups, moreover, has endured and exerts an overriding significance in explaining late twentieth-century politics in Guyana, Suriname, and Trinidad (Chapter 8).

The first imported Indians inhabited wooden barracks or "ranges." Many received such harsh treatment that the British Anti-Slavery Society asserted that slavery had been reintroduced in a new guise. As indenture terms expired, Indian laborers often chose to take up small parcels of local lands rather than returning to India, an option encouraged by planters who continued to rely on new Indian settlements as reservoirs of estate labor. By the end of the nineteenth century, former indentured plantation workers from India were thereby establishing their own village communities. And they already were being identified with local Caribbean livelihood activities that have continued to characterize those groups into the late twentieth century: rice production in Guyana, sugar-cane farming in Trinidad, and small-scale cultivation in Guadeloupe and Suriname.

India was not the sole source of labor introduced into the Caribbean after slavery. Contingents of free Africans "liberated" from Spanish and Portuguese slave ships, were brought to the British islands after 1838. A variety of Europeans, notably Portuguese Madeirans, came in the middle decades of the nineteenth century. Yucatecan Indians were "deported" to Cuba after the Caste Wars between the Maya and Mexicans at mid-nineteenth century. In the last decades of the century, tens of thousands of sugar-cane workers from Spain, mainly from Galicia and also from the Canary Islands, augmented the Cuban plantation labor force. Between 1848 and 1874, 125,000 Chinese were brought to Cuba: "Once landed, they were offered for sale as though they were slaves, although technically it was their contracts that were sold" (Scott 1985: 29). Chinese contract workers toiled alongside black slaves in Cuban sugar-cane fields, and their treatment and housing hardly distinguished them from their enslaved coworkers. All of these peoples – Asians, Europeans, and Africans introduced into the Caribbean region after slavery as plantation workers – lived in crude housing, labored long hours, and were treated with the same rough anonymity that always had been accorded the region's black slaves.

Nearly four centuries after the remarkably persistent institution of Caribbean slavery had been implanted in the Greater Antilles, it came to an end. Cuban planters favored a "gradual" end to slavery, thereby avoiding immediate freedom. The result was the so-called *patronato* arrangement passed into law early in 1880 providing Cuban planters with continued labor and ex-slaves with "tutelage," an institution roughly analogous to the apprenticeship system of the British islands half a century earlier. One stipulation of the Cuban law was that individual *patrocinados* would be freed if former masters violated the statute's provisions, such as withholding pay. Ex-slaves continuously obtained their freedom in this manner until the *patronato*, and Caribbean slavery, finally came to an end in 1886 (Scott 1985: 40, 127–71).

One reason slavery persisted in the Spanish Caribbean was political. The issue of emancipation had been intertwined with Cuba's Ten Years War of 1868–78 in which both the insurgents fighting for independence and Spain had included in their manifestos limited forms of freedom for slaves. But politics and economics can never be separated; even though Spain formally had proposed freedom for children born of Cuban slaves after 1868, this proposal still was strenuously opposed by a Spanish merchant class who feared the subsequent unraveling of the Cuban sugar industry and an associated disintegration of their profitable Cuban market (Scott 1985: 64). Similarly, in Puerto Rico (where slavery did not end until 1873) the "Golden Age" of sugar-cane development early in the nineteenth century involved relatively few slaves. But economic imperatives influenced harsh legal stipulations in Puerto Rico, binding landless freedmen to particular sugar-cane plantations in order to assure an adequate pool of estate labor (Mintz 1974b: 91–94).

In the latter half of the nineteenth century the people of Cuban sugar-cane plantations were linked closely to international information flows that provided news about sugar technology, markets, transportation improvements, and much else. This kind of interrelationship with external places and peoples had been a key ingredient of the Caribbean plantation since its inception. But by the late 1800s the Cubans were receiving and sending out goods, people, products, and news in a continuous circulation not only to Spain but to neighboring colonies and states throughout the Americas. These information flows, significantly, were not limited to the Cuban elite but included slave men and women as well. During the 1860s, for example, the words of a song said to be sung among Cuban slaves were reported as "advance, Lincoln, advance. You are our hope" (Scott 1985: 38). It is indeed ironic that Cuban field slaves sought spiritual inspiration from concurrent events in the United States. In the ensuing decades those same Cuban field slaves and their descendants would have opportunities to respond to stimuli emanating from the United States in a much more direct way.

# 4

## The American century

The Monroe Doctrine, so far as the Caribbean is concerned, is no
longer a policy as much in the interest of other American republics
as our own. We proclaim the scarcely veiled doctrine of "special
interest." We profess to seek no territory, but we maintain the
supreme importance of our own interests in the Caribbean – in a
word our hegemony – against everyone else, including the peoples
of that region themselves.

Leland Hamilton Jenks, *Our Cuban Colony*, 1928.

The interest in the United States expressed by the Cuban slaves of
the mid-nineteenth century was reciprocated by the interest that US
slaveholders had in Cuba. To some Americans, Manifest Destiny
meant moving not only west but also south (May 1973). In 1854
United States President Franklin Pierce authorized an emissary to
offer Spain $130 million for Cuba, resulting in the famous Ostend
(Belgium) Manifesto that declared the geopolitical inevitability of
the US domination of Cuba. Some expansionists went even further,
proposing an eventual "golden circle" of US slave states along the
Mexican Gulf Coast, Central America, and then north through the
Antilles, forming altogether an economic colossus of subtropical
agricultural production (McPherson 1988: 103–16).

United States-Caribbean economic ties had, of course, predated
US political independence, Yankee traders often penetrating illegally
the controlled commerce between European nations and their Caribbean colonies. Indeed, the first external recognition given to the
proclaimed political independence of the United States is said to
have been an eleven-gun salute off Dutch St. Eustatius in November,
1776 (Tuchman 1988). Links were particularly strong to the British
Caribbean in the United States' formative years: Alexander Hamilton, the first US Secretary of the Treasury, was born on British
Nevis; and William Thornton, a planter from tiny Jost Van Dyke in
the British Virgin Islands, designed the US Capitol building as well

78

as several of the original buildings at the University of Virginia (Williams 1970: 241–42). Prior to the end of the eighteenth century, US traders had begun to supply massive amounts of North American flour and other foodstuffs to Spanish Cuba, a commercial connection leading to contemporaneous American investments in Cuban sugar cane (Liss 1983: 113).

US enthusiasm about annexing Caribbean territories at mid-nineteenth century was by no means a whimsical notion. President James Buchanan, following Pierce's lead, urged negotiations with Spain for Cuba in the late 1850s (McPherson 1988: 194–95), and President U. S. Grant favored the annexation of the Dominican Republic two decades later. But the United States had just finished a convulsive war centered on the issues of slave production and slavery itself. Anti-imperialist sentiment, furthermore, countered expansionist feelings with, among other arguments, a curiously environmental deterministic line of reasoning: warm climates induced sloth and laziness among a resident populace, so the inevitable expansion of the United States into the Caribbean might better be commercial, it was argued, rather than political because the latter arrangement would involve attendant social responsibilities for entire populations infected with tropical indolence (Healy 1970: 214–15).

The eventual commercial expansion by the United States into the Caribbean as the twentieth century began was part of a global imperialism: European nations – particularly Britain, France, and Germany – recently had annexed enormous chunks of the Eastern Hemisphere as colonial territories, principally areas of Africa and Southeast Asia that had not yet been colonized by Europe. And though many Americans considered Europe's colonial expansion vulgar and distasteful, United States' indifference toward developments in her own Caribbean "backyard" could easily lead to a European neocolonialism there that would reduce the buffering effect that the Atlantic Ocean always had provided for Americans. In any case, after the brief Spanish–American War of 1898, the United States stood alone as the supreme imperial power in the Caribbean region, the presence of a number of small European colonies there notwithstanding. The war also may be said to have marked the beginning of a new United States imperialism, whether Americans liked it or not, that saw an expansion of US power not only into the Caribbean but also into the Pacific, notably with the American takeover of the Philippines from Spain.

At a global scale, American expansion into the Caribbean helped to mark a crucial geopolitical watershed. In some ways it was part of British accommodation with former rivals in other parts of the world, not only the Americans in the Caribbean but also the Japanese in the northern Pacific and the French in the Mediterranean, in order to allow Britain to focus on the growing German threat in Europe (Gilpin 1981: 196–97). It also inspired the United States to buttress its commercial interests in the Caribbean with military force, in part because of aggressive German imperial designs for parts of the Caribbean and Latin America (Collin 1990: 69–70).

Unlike European nations, the United States usually did not take direct political control of Caribbean territory they dominated; the exceptions were Puerto Rico and the former Danish Virgin Islands of St. Croix, St. John, and St. Thomas. On Puerto Rico the US military occupation in 1898 was followed in 1900 by the Foraker Act passed by the US Congress. The law terminated the military government, supplanting it with a civilian regime whose key officials were to be appointed directly by the US President (Silvestrini 1989). The US purchased the Virgin Islands from Denmark in 1917 for US $25 million, mainly for the harbor facilities at St. Thomas.

Yet US domination of the entire Caribbean region came to be conducted as if the region's several independent states were American political colonies. Americans made many and varied demands on their host Caribbean countries in the first decades of the twentieth century, particularly for "free access to resources, favorable market conditions, a docile working class, a compliant political elite, and a friendly climate of investment" (Pérez 1986: 110–11). In world-system parlance, the Caribbean region had by no means "disintegrated" as a key peripheral zone under the onslaught of the European sugar beet (Taylor 1988). Rather, the Caribbean region's peripheral status had been reinforced as it became an economic appendage to the newly emerging core economy of the United States (Agnew 1987: 13–15).

The events of the 1890s that paved the way for the unfolding of the decades thereafter suggest that the twentieth century in the Caribbean might be called "the American century". Despite the influences from Europe, Canada, and neighboring Latin American countries, the United States has been by far the dominant geopolitical power in the Caribbean in this century. The physical juxtaposition of tiny Caribbean states and the United States during this century, further, helps to explain much about the Caribbean at all levels. Europe's first overseas colonial outpost, the Caribbean might also be called the original US-dominated world region.

## The Caribbean rim of Central America

The enduring event symbolizing US domination over the entire circum-Caribbean early in the twentieth century was the construction of the Panama Canal from 1904 to 1914. Realizing a dream as old as the Spanish conquistadores of a trans-Isthmian canal, US engineers earlier had considered Nicaragua a possible canal route, an idea that fell from favor in the light of Mt. Pelée's explosion in Martinique in 1902 accompanied by reports of simultaneous volcanic activity in Nicaragua. One year later, in November 1903, the Republic of Panama (formerly a province of Colombia) was born with considerable US help: US gunboats, following a perverse interpretation of a US treaty with Colombia to "protect" the Isthmus from invasion by a foreign power, protected a Panamanian "revolution" from Colombia herself (McCullough 1977: 377–79)! President Theodore Roosevelt, facing opposition at home for such transparent and heavy-handed action, explained to the US Congress early in 1904 that the United States owed it to civilization to construct the Panama Canal. And later in that year he pointed out to a British diplomat that a benign but protective United States now would play a commanding role within the Caribbean region: "The people of the United States and the people of the Isthmus and the rest of mankind will all be the better because we dig the Panama Canal and keep order in its neighborhood" (Williams 1970: 422).

The construction itself put the world, not simply the Caribbean or the rest of the hemisphere, on notice that US technological supremacy could solve nearly any problem imaginable. The French, after all, had attempted the creation of a canal through Panama in the 1880s and had failed. In contrast, the US construction effort was indeed of heroic scope. It involved blasting through a giant mountain ridge at Culebra, forming an artificial lake 85 feet above sea level, and building massive sets of locks at either end of the canal to lift ocean-going vessels up to lake level and then down again. This supreme engineering feat was accompanied by other superlative achievements, notably the control of yellow fever that had helped doom the French attempt two decades earlier.

The US canal construction was an accomplishment shared by tens of thousands of black laborers from the insular Caribbean, notably those from Jamaica and Barbados but also from other British, French, Dutch, and Danish islands (Chapter 6). Some of the laborers drifted west after the canal project was finished. Many located jobs along the eastern rim of Central America in the bananalands established there by the American-owned United Fruit Company at the turn of the century.

Vast areas of the coastal banana zones along the Caribbean rim of Central America had been obtained by the US fruit companies for low prices, or sometimes nothing, in return for railroad-induced "development" of the region. During the 1920s, after the infant industries had gained major footholds and a widespread banana market established itself in the eastern United States, local Central American governments stiffened markedly. They sought more taxes from their corporate guests, overtures often rejected by company officials claiming tax exemptions. Then, during the depression decade of the 1930s, Central American politicians – bedeviled by unemployment and political unrest – found themselves in weak positions to combat US banana interests that threatened, if pushed, to create even more unemployment. The power exerted by the United Fruit Company in Central American affairs by 1935 was awesome: "The corporation . . . [that] monopolizes the banana trade . . . [and] owns or controls railroads, docks, steamships, radio, housing facilities, leading wholesale and retail stores . . . and which also controls the livelihood of many business men, farmers, laborers and professional men, can speak with such force that politicians accede to its will" (Kepner and Soothill 1935: 241).

The contrasts in housing, pay, and work conditions for white American overseers and their black West Indian laborers in both the Canal Zone and also in the isolated banana company compounds were ugly testimonies to the racism of the period. In Panama, "gold" (white staff) and "silver" (black laborers) payrolls were the euphemisms for racial segregation in housing, schools, pay levels, and canteens. The perpetuation of racial segregation in the Canal Zone, moreover, lasted well into the 1950s (Conniff 1985). The stark contrasts in housing quality in the Canal Zone were duplicated in the Central American banana enclaves where airy, screened bungalows occupied by white Americans were placed conveniently out of sight from the unhygienic wooden shanties of the workers, many of which had dirt floors. In theory, small-scale banana producers had the same access to port facilities as did the giant corporations, although the company's banana inspectors routinely rejected small producers' crops for shipment, especially when the market was weak. Subsistence production among the banana workers was not encouraged, and most bought their food at company stores, an arrangement creating undue hardship during periods of low banana prices and therefore low wages (Kepner and Soothill 1935: 28–30, 265–71, 319).

## Domination of the Greater Antilles

The United States' effort in building the Panama Canal was, naturally enough, accompanied by a regional military presence intended to secure the Isthmus and the sea lanes leading to it. So Theodore Roosevelt's famous "big stick" policy was entirely visible in the waters of the Caribbean in the first decade of the twentieth century. In 1905 the larger units of Britain's West India squadron withdrew from the region, leaving the US Navy supreme from Miami to Yucatán to Trinidad. And, from 1900 to 1930, the slightest indication of political unrest or civil disturbance within the circum-Caribbean zone was answered, as often as not, by the appearance of a US naval vessel hovering offshore. The Greater Antilles were seen as particularly crucial in protecting routes to the canal, especially against the German threat during World War I. So, by the 1920s, US troops had occupied – at least for brief periods – Cuba, Haiti, the Dominican Republic and, of course, the new American territory of Puerto Rico.

But economic penetration had preceded a US military presence in the Greater Antilles. As early as 1851 the Consul General of Cuba in the United States described his country as a political colony of Spain but an economic colony of the United States (Ortiz 1947: 64). By the 1870s, Cuba, Puerto Rico, and the Dominican Republic were shipping virtually all of their sugar to the east coast of the United States aboard US vessels. And American sugar refiners, by the early 1880s, had begun to popularize a high-grade granulated sugar in standardized packages. These innovations helped to heighten consumer demand in the United States and intensified production in the Greater Antilles despite competition from US sugar. By the end of the nineteenth century highly refined sugar could be stored for long periods, thereby allowing importers to stock surpluses and then to pressure non-US producers to lower prices. Improvements in refining standards and storage techniques were, moreover, elements in radically new developments in the world sugar industry in the last decades of the 1800s that also featured giant monopolistic corporations, faster product transportation, and transmissions of price and market information relayed by the international telegraph system (Jenks 1928: 28–29; Moreno Fraginals 1985).

In February, 1895, at the height of the sugar-cane harvest, a war against Spain broke out in Cuba. For the next three years, and at a cost of an estimated 400,000 Cuban lives, the Cuban rebels gained control of most of the countryside despite particularly cruel Spanish tactics featuring concentration camps and forcible relocations of segments of the rural populace (Chapter 7). In 1898, United States forces entered the war on the side of the rebels. The brief Spanish–American War then secured Cuban independence from Spain, and it also resulted in Spain ceding Puerto Rico, the Philippines, and Guam to the United States. Dr. Andrew S. Draper, President of the University of Illinois, possibly exemplified general US opinion toward the war as an "episode in . . . [a] . . . world-wide contest for self-government" and typical of the "unselfish, neighborly, and resolute spirit shown by the United States toward its weaker sister states of the Caribbean and the Pacific" (1899: 6). Cuban nationalists interpreted US actions in a different way. The United States failed to recognize the Cuban revolutionary assembly and also prevented this grassroots body from participating in the eventual Spanish surrender. In brief, the Cuban revolutionaries saw the US entry as redirecting the war effort away from true Cuban independence and toward a neocolonial identity: "A Cuban war of liberation was transformed into a North American war of conquest" (Pérez 1986: 30).

After Spain's withdrawal from Cuba, a US military government concentrated on rebel disarmament, civilian food relief, and disease control. In 1902 the US military occupation ended (US marines would be back from 1906–9 and again from 1917–22) with the formation of the Republic of Cuba. The new constitution, however, was as much a creation of Washington as it was of Havana. In particular, the eight articles of the Cuban constitution's Platt Amendment provided the political justification for "US capital penetration and the appropriation of local resources" (Pérez 1986: 138) throughout the first half of the century. Among other things, the Platt Amendment recognized Washington's "right to intervene for the preservation of Cuban independence" and called for the establishment of US naval bases on Cuban soil, notably at Guantanamo Bay on the extreme southeastern end of the island (Jenks 1928: 78–84).

Under US economic control in the first decades of the twentieth century, Cuba accelerated her pace of agricultural development. In the last years of Spanish rule transportation and refining improvements for Cuban cane sugar had been paralleled by the replacement of the many small mills of the past with new, enlarged grinding factories: "By 1890, Cuba's famous *central* Constancia was producing 135,000 sacks of sugar per season – the biggest sugar mill in the world" (Mintz 1964: xxviii). Yet the tempo of modernization quickened even further in response to US investments after the Spanish–American war. Stimulated by a 20 percent reduction in the US tariff on Cuban sugar, American private investment in Cuba soared from $220 million in 1912 to $1.325 billion in 1928 (Guerra y Sánchez 1964: 70–71; Kepner and Soothill 1935: 20). Then, as sugar prices fell in the early 1920s and Cuban landowners were forced to sell out at low prices, Americans came to dominate Cuban land and milling directly; more than 60 percent of the Cuban sugar-cane crop of 1926–27 was processed at American-owned mills (Jenks 1928: 284). But the Cuban sugar industry was not the only sector of the country's economy coming under US control. By 1928 new foreign factories – including Mennen, Armour, Proctor and Gamble, Colgate, Pabst, and Fleischmann from the United States – had entered the Cuban light-manufacturing and food-processing industries, thereby inhibiting indigenous Cuban industrial develop-ment (Pérez 1986: 336).

Transformations of Cuba's cultural landscape in the first decades of the twentieth century reflected the harnessing of Cuban sugar-cane production to the US consumer market. Cuban railroad con-struction, extended in the late 1800s and facilitated by new laws passed during the US military occupation, created great "tentacles" of steel tracks that "turned the centrals into monstrous iron octo-puses" whose voracious appetites called for ever more sugar-cane planting (Ortiz 1947: 51–52). The spread of large-scale cane plant-ings, furthermore, undermined the traditional small-scale *colono* farmer of the Cuban countryside whose self-sufficiency then became wage dependency as he joined the growing throng of seasonal cane cutters and millhands. The drifting army of sugar-cane workers, whose activities were tied to the rhythms of the *zafra* (harvest) and *tiempo muerto* (the "dead time" when the cane was growing), found themselves competing for wages with introduced laborers in their own country. In 1921 25,000 migrant laborers from Jamaica and Haiti came to work in the Cuban cane harvest, creating the overall effect of driving wages down (Guerra y Sánchez 1964: 145).

Similar changes altered the face of Puerto Rico owing to the US annexation of the island: sugar-cane acreage, which already had expanded dramatically under Spanish control in the late nineteenth century, more than trebled between 1899 and 1940. This expansion, not surprisingly, was linked closely to the establishment of large, US-owned grinding mills, the semi-arid south coast of the island becoming a particularly important irrigated sugar-cane zone. The traditional independent *jíbaro* countryside farmer of Puerto Rico, like the *colono* of Cuba, found himself involved in a money economy more than ever before as expanding cash crops outcompeted local food crops and livestock. Puerto Rican tobacco cultivation, mainly in the east central part of the island, also thrived because of the availability of the US market although, unlike sugar cane, the intensive, small-scale production of tobacco remained mainly in Puerto Rican hands. The heightened export market for agricultural staples and the related need for the importation of more and more foodstuffs led to occupational complexity in Puerto Rico with the emergence of local merchants, shippers, and handlers. Similarly, the dispensation of US government services (Puerto Ricans were granted US citizenship in 1917) necessitated an increase of civil servants and teachers. New members of a Puerto Rican middle class resided mainly in the new urban areas, and the population of greater San Juan grew to 400,000 by 1950 (Steward *et al.* 1956: 70).

Critics of the US takeover of Puerto Rico seized mainly on the sugar industry that concentrated land in fewer hands, tied formerly self-sufficient farmers to wage labor and, in some cases, intervened in the social and civic life of the working class (Steward *et al.* 1956: 70). American investors in Puerto Rican sugar companies realized sizable profits in the first decades of the twentieth century. In most years investment yields ranged in the 10 percent range, and in the boom period immediately following World War I, stock dividends of 60 and even 70 percent were reported for Central Aguirre, one of the giant Puerto Rican sugar mills (Williams 1970: 436).

The overwhelming presence of US sugar interests in the Greater Antilles in the early twentieth century led not only to a massive takeover of peoples and lands, but it also transcended economically the constraints of insularity. Besides owning land and factories, American sugar companies regulated seasonal labor flows, built highways, railroads, warehouses, and wharves, and came to own electrical generating plants and animal herds. Company steamers hauled raw sugar to the eastern US for final refining. On occasion sugar company lighters even carried cut canes from the Dominican Republic across the Mona Passage to Puerto Rico for grinding, thereby achieving an interisland integration of intimately related economic activities that traditionally had been confined to individual Caribbean estates (Ortiz 1947: 51–52).

US interests in the Dominican Republic had been solidified by a treaty between the two countries in 1891, followed by the establishment of the Santo Domingo Improvement Company (based in New York) the next year. The company controlled, among other things, part of the rail network as well as the national bank of the Dominican Republic (Munro 1964: 17). Soon thereafter, intensive US-controlled sugar-cane production began to dominate the Dominican countryside, again reducing small landowners to seasonally employed estate workers. In 1921 directors of the (American) Central Romana estate literally burned to the ground two tiny agricultural villages in the southeastern part of the Dominican Republic that were standing in the way of estate sugar-cane expansion (Calder 1984: 91, 98). As in Cuba and Puerto Rico, sugar-cane production in the Dominican Republic called for massive imports of food and light-manufactured items formerly produced locally. In 1921, the United States accounted for 84 percent of all imports into the Dominican Republic, from food and lumber to machinery and automobiles (Knight 1928: 149–50).

American economic domination of the Dominican Republic was paralleled by military intervention; US troops occupied the country from 1916 to 1924. From 1911, when Dominican president Ramón Cáceres was assassinated, to 1916, US warships had kept a close watch on the country in order to intimidate possible revolutionaries. The military occupation itself was marked by American attempts to improve education, sanitation, and public works (roads, sewers, bridges). There were also attempts at agricultural extension work and prison reform. The land law of the Dominican Republic was revamped by the US military government, paving the way, incidentally, for greater US control of Dominican lands. Doubtless the most damaging legacy of the US occupation was the training of the Dominican army-police force (the *Guardia Nacional Dominicana*). By 1926, a local leader of the Guardia, Rafael Trujillo, used the force to reduce local political competition. Trujillo gained control of the country in 1930, using the US-trained military as a springboard to rule the Dominican Republic with incredible corruption and brutality until 1961 (Calder 1984: 61).

US military occupation of Haiti (1915–34) by marines and naval troops overlapped the occupation of the Dominican Republic. American interest in Haiti was not focused on large-scale sugar production as it was elsewhere in the Greater Antilles. Rather, it was based more on geopolitical concerns, specificially the fear of Haitian anarchy and its possible exploitation by competing metropolitan powers, mainly Germany (Plummer 1988). Shortly after intervention, American naval officers began to supervise the collection of all customs duties through the American-controlled Banque Nationale, actions that effectively dominated Haitian internal finances. American control over Haitian land was less successful. A new, American-sponsored constitution in 1918 featured an "alien landownership" clause designed to facilitate American capital investment, but the Haitian peasantry was loathe to sell out their lands, and US officials were reluctant to force the issue (Schmidt 1971: 69–70, 113, 179).

Political turmoil in Haiti had alerted United States forces for possible landing operations in Haiti as early as July, 1914. Then the grisly atrocities perpetrated by Haitian president Vilbrun Guillaume Sam toward his political opponents, followed by Sam's public execution in July, 1915, brought Rear Admiral William B. Caperton to lead forces of the US Navy to occupy Port-au-Prince and then to govern Haiti for a year. American newspapers, typified by the *Chicago Tribune* which described Haiti as "a rebellion called a republic," satisfied themselves that the invasion was necessary for a neighbor state unable to govern itself (Healy 1976: 123).

The two decades of US Marine occupation satisfied none of the diverse elements of Haitian society. The brown-skinned Haitian aristocracy considered the marines rude and boorish, but the US troops encountered their stiffest resistance from the black peasantry of the countryside. Especially in areas where the marines employed rough, corvée labor recruiting tactics for road construction, the Haitian *caco* irregulars responded with hit-and-run raids against marine patrols (Chapter 7). A nationwide strike and a series of riots in October and November, 1929, caused ostensibly by low coffee prices but later focused on the presence of American troops, was a catalyst for eventual American military withdrawal from Haiti in 1934 (Schmidt 1971: 104–5, 196–207).

The American-financed expansion of the sugar industry in the Greater Antilles helped to create the suffering felt there during the world economic depression of the 1930s. Starting in 1925 the Cuban sugar industry began to decline owing to lower prices (Langley 1982: 138). But subsistence alternatives for local workers had been reduced since 1900 because much land formerly devoted to food crops had been planted in sugar cane, taken over by American investors, or both. New sugar treaties between the United States and Cuba in the 1930s then intensified even greater Cuban dependence on US sugar quotas; Cuban sugar exports to the US rose by 40 percent between 1933 and 1936 while US sales in Cuba were increasing by 140 percent (Heston 1987: 392–93). The effects of the price decline for Greater Antillean sugar, furthermore, were felt throughout the entire Caribbean region.

In 1929 the Dominican Republic sharply reduced the number of seasonal migrant workers from the British islands of the Lesser Antilles who had been coming to harvest cane for two decades. Cuba prohibited Haitian and Jamaican workers in 1933. A series of riots followed throughout the small British islands during the 1930s (Chapter 7); they were caused in large part by the curtailed emigration outlets for migrating men who had come to depend on earning US dollars in Cuba or the Dominican Republic seasonally to supplement depression-reduced livelihood possibilities on their home islands (Richardson 1983: 140–44).

A strike in Havana in August, 1933 (during the *tiempo muerto* when Cuban strikes and labor disturbances were always prevalent) led to the resignation of the dictator Gerardo Machado, a strong supporter of US business interests. An interim government led by Dr Ramon Grau, who unilaterally annulled the Platt Amendment in 1933, had partial support from increasingly restive Cuban students. In 1934, a pro-US government was restored with the backing of Fulgencio Batista y Zalvidar, a leader of the "sergeants' faction" of the Cuban army. Batista himself became the Cuban president in 1940 (Langley 1982: 141–46).

On the eve of the US involvement in World War II, the United States arranged to supply Britain with fifty near-obsolete destroyers in return for 99-year leases of naval bases in Antigua, the Bahamas, British Guiana, Jamaica, St. Lucia, and Trinidad and also in Bermuda and Newfoundland (Williams 1970: 426–27). When the war began the United States Congress reaffirmed a No-Transfer Resolution that originally had been a corollary of the Monroe Doctrine. It meant that no European country could transfer Western Hemisphere colonies to another country without US approval, and it was directed toward the Dutch and French Caribbean colonies whose metropolitan countries had come under German control. American anxiety was focused particularly on the oil refineries of Aruba and Curaçao where Venezuelan oil was refined and on the bauxite (aluminum ore) deposits from Suriname. German submarines were indeed active, especially during 1942, when more than 300 ships in Caribbean waters were damaged or destroyed. American anti-submarine patrols thereafter reduced the effectiveness of attacks on vessels carrying raw materials northward.

The overall effect of US involvement in the Caribbean during the war was to tie the region ever closer to the United States. European markets had become disrupted, and now even the small islands that traditionally had looked to Britain, France, and the Netherlands for food supplements and manufactured goods, were turning more than ever to the United States. Many of the islands' peoples, moreover, were becoming accustomed to working for the "Yankee dollar" rather than for pounds, guilders, or francs (Prest 1948). The shift toward the United States continued to reinforce the Caribbean's colonial status because "in a significant way it . . . retarded Caribbean economic diversification by preserving the area's monocultural character" (Langley 1982: 180–81).

## Revolutionary Cuba and the United States

When Fidel Castro and his small band of guerrilla soldiers entered Havana in January, 1959 (Chapter 7), in the wake of President Batista's flight, the immediate US reaction was perhaps one of surprise more than anything else (Francis 1967). Most Americans already knew of Castro and his romantic forays from mountain strongholds against the corrupt Batista regime, yet the suddenness of Batista's collapse was unexpected. But surprise soon turned to suspicion. In the months following, reports of public executions of former Batista henchmen and Castro's anti-US rhetoric transformed the American image of the revolutionary Cuban leader from that of a slightly amusing bearded Latin with a penchant for cigars and baggy fatigue uniforms; soon he was seen as an ungrateful menace whose fiery rhetoric – emitted barely 90 miles from the southern tip of Florida – might infect neighboring Caribbean islands and the rest of Latin America, thereby extending Communism throughout the hemisphere from the very doorstep of the United States.

Cuban bitterness toward the United States, deep-seated resentments successfully tapped by Castro, was entirely comprehensible from a Cuban point of view. The historical economic factors were reasons enough for strong anti-US feelings: the earlier thwarted revolution against Spain, Cuban dependence on US markets for its agriculture and mineral production, the misery of the seasonal *tiempo muerto* related directly to Cuba's fatal reliance on the US sugar quota, the lavish dwellings of the few rich Cubans compared with the country's grinding rural poverty. And all of these issues were analyzed and dissected by vocal and increasingly militant university students. Yet these issues were abstract compared with the grassroots animosity that had built up against the United States and its citizens in the decades prior to revolution. Thousands of

American tourists visited Cuba in all seasons of the year, and the island became well-known as a naughty playground featuring vice and gambling, activities enriching gangsters and corrupt local officials. Ordinary Cubans were humiliated and ashamed. Slot machines were everywhere, their profits raked off by government officers. Worst of all was the prostitution: an estimated 270 Cuban brothels housed prostitutes, some as young as 12; prostitution was centered in Havana, but the poverty of the Cuban countryside forced thousands of young Cuban women into the city's brothels and bars that catered to Americans visiting Havana on week-end excursions (Mills 1960).

The understandable Cuban resentment, from both immediate and long-term events, helps to explain the sudden Cuban political shift away from the United States and toward the Soviet Union in the months following Batista's overthrow, although observers then and since are by no means agreed as to exactly why US–Cuban relations soured so quickly. While battling Batista, Castro initially had identified himself as anti-imperalistic but definitely not anti-American (Langley 1982: 212–13). But the portrayal of post-revolution trials and executions in Havana as vengeful and bloody circuses in US newspapers and magazines angered Castro into incendiary anti-US tirades within a month of his takeover (Matthews 1975: 134; Robbins 1985: 82). The January 26, 1959, cover of *Time* featured Castro's face towering above a mountainous tropical landscape that was literally ablaze. The story inside described the holding of courts-martial in a park in downtown Havana as "an unlikely spot for cool justice but perfect for a modern-day Madame Defarge." Immediate Cuban economic reforms involved the revolutionary government's "intervention" in Cuban Telephone (an American-owned company) in March, 1959, followed by the nationalization of American landholdings, refineries, and factories, takeovers that provoked US wrath (Langley 1982: 216–18). Castro himself may have realized from the start that his revolution – beginning in Cuba but inevitably extending itself to other Latin American and Third World areas – was on an eventual collision course with US interests (Smith 1987).

Whatever the reasons, it took only two years after the Batista overthrow for the US to sever diplomatic relations. Cuba established diplomatic ties with the USSR in May, 1960. That summer Cuba received copious quantities of small arms from Czechoslovakia. In July, 1960, the US Congress terminated the Cuban sugar quota. Castro ordered the nationalization of the sugar industry in October. In early January, 1961, the Cuban government banished nearly all

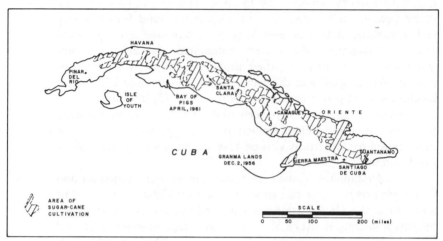

Map 3   Cuba

US embassy personnel from Havana. On January 4, 1961, the United States broke formal relations with Cuba and began a concerted campaign among states of the Western Hemisphere to have Cuba branded as an outcast country.

But the United States was unaccustomed to relying only on diplomacy in order to have its own way in the Caribbean region. Simultaneously with Castro's emergence in Cuba were American plans to intervene militarily, not directly with American soldiers but by using dissident anti-Castro Cubans. By Autumn, 1960, the US Central Intelligence Agency was training a Cuban "brigade" in Guatemala for a possible invasion of Cuba. The CIA training and plans were, incredibly, so well-known that US newspapers carried routine reports enumerating men and material (Langley 1982: 223–25). President John F. Kennedy, inaugurated in January, 1961, complained that all the Cubans needed to monitor US preparations for invasion were newspaper subscriptions (Higgins 1987: 117).

The ill-fated Bay of Pigs operation was launched from Puerto Cabezas, Nicaragua, on April 14, 1961. The roughly 1,400 men of the invasion force were said to be from among the Cuban upper and middle classes, soldiers whose personal interests in their own land-holdings and businesses in Cuba, according to Castro, far outweighed their collective zeal for Cuban democracy (Matthews 1975: 202). The operation itself never had a chance: the half-hearted and hesitant planning deprived the invaders of effective air support; and the main invasion party encountered unexpected coral reefs as well as active resistance from Cuban military and civilian units, all of whom, the invaders had been assured, would immediately join them to topple the Castro government. The end result of the Bay of Pigs, a term that has become synonymous with blundering ineptitude, was "simply to drive Cuba into the arms of the Soviet Union as a nearby military or even nuclear base against the United States" (Higgins 1987: 176).

In the following year, US overflights of Cuba produced photographs of missile launch pads and the installation of two kinds of Soviet missiles on Cuban soil, both said to have a "first strike" capacity against the United States and capable of reaching targets nearly as far away as Canada (Abel 1968: 58–59). The ensuing Cuban missile crisis of October, 1962, in which the Soviets agreed to remove and take back the weapons, was reciprocated by the American dismantling of outmoded missiles in Turkey and a pledge not to invade Cuba, the latter considered a major concession from the Soviet and Cuban viewpoints (Garthoff 1987: 58). Most importantly, the 1962 missile crisis – played out in a Caribbean locale – is widely regarded as the closest the world has ever come to nuclear war. And a remarkably frank dialogue between American and Soviet protagonists occurring nearly three decades after the crisis, has thrown new light on the frightening events of October, 1962 (Blight and Welch 1989; Keller 1989).

Within Cuba itself, the unsteady drift away from capitalism was not facilitated by either the inexperience of the revolutionary administration or the continuous need for war mobilization. The new government soon eliminated most institutions that had existed under Batista and substituted new economic and social organizations. Agrarian reform laws in 1959 and 1963 reduced the size of private landholdings and emphasized state and collective farms, transformations implemented by the INRA (*Instituto Nacional de Reforma Agraria*), a government agency that came virtually to run the entire Cuban countryside. The new regime emphasized the rural areas at Havana's expense, building new schools, factories, and health clinics in the towns, villages, and cane lands that had been so neglected prior to 1959. The historic shortfall of the 1970 Cuban sugar-cane harvest seems to have marked an economic watershed for the country. The disappointing results of that year's crop led to severe deprivations in basic consumer goods and food items, but after 1970 productivity increased owing to an overall improvement in fields, factories, and transportation facilities. Perhaps the most notable local success of the Castro government was in the area of personal health. Malaria, diptheria, and typhoid were eliminated, or nearly so, by 1971. And, by the mid-1970s, Cuba had the lowest infant mortality rate (27.7 per 1,000 births) in Latin America (Matthews 1975: 364–65).

Despite US attempts to isolate Cuba, she had established diplomatic relations with several neighboring circum-Caribbean states by 1975. In 1977, the United States and Cuba opened reciprocal "interests sections" in Havana and Washington (Langley 1982: 276). Then, during 1979 and 1980, thousands of former Cubans who had earlier fled to the United States were allowed to visit friends and families in Cuba. A series of small-boat hijackings by Cubans desirous of heading for Florida paralleled the visits. The resultant unrest, blamed by Cuba on deliberately slow processing of entry visas by US immigration officials, was a prelude to the so-called "boatlift" from the coastal fishing town of Mariel, west of Havana. Cuban emigration restrictions suddenly were lifted on April 21, 1980. From April to September, 1980, an estimated 120,000 Cubans emigrated freely to the United States. The social cost was high for both countries: those opting to leave Cuba were branded as *gusanos* ("worms") and some were beaten and harassed, marking an ugly and enervating summer in Cuba during which little work was accomplished; the United States, on the receiving end of the flotilla of Cubans sailing from Mariel, was flooded with immigrants, a number of whom were found to be habitual criminals or mentally ill (Smith 1987).

## Military intervention in the Dominican Republic

The American government's fear of "another Cuba" helps to explain the massive US military intervention in the Dominican Republic only six years after Castro's ascent to power. A steadily deteriorating political situation in the eastern half of Hispaniola ended abruptly with the existing government's collapse in late April, 1965. Rival political groups then clashed, and several hundred Dominicans were killed. Amidst a chaotic series of events, and acting mainly on rumor and confusion, US decision-makers dispatched a marine battalion to protect American lives and property. Then, as conditions worsened, the United States descended on the Dominican Republic with soldiers, vehicles, and weaponry. Before mid-May 1965, the US had airlifted to Hispaniola 23,000 troops, nearly half the number serving in Vietnam at the time (Lowenthal 1972: 111–12). Whereas Cuba's guerrilla insurgency was restricted to the countryside, the sporadic clashes in the Dominican Republic – and therefore the presence of most of the introduced US soldiers – were limited mainly to the urban area of Santo Domingo. The US military occupation of the Dominican Republic lasted for a year, brought about an uneasy peace, and also encouraged right-wing political regimes throughout the Western Hemisphere.

The assassination of the Dominican dictator, Rafael Trujillo, in May, 1961, had been followed by the election in 1962 of Juan Bosch as the country's president. A tireless campaigner who had built his victorious Dominican Revolutionary Party from among the ranks of the working classes, Bosch soon attempted to break up the vast Trujillo family holdings for the sake of small landholders and to reduce the power of both the military and the Catholic church in Dominican affairs. In September, 1963, Bosch was ousted by a military coup because of his suspected Communist sympathies and was replaced by a civilian council whose real power lay in its military backing. The following eighteen months were exceptionally difficult because the worsening political scenario was paralleled by a downward economic spiral caused by severe local drought conditions, low prices for Dominican agricultural exports, and a dock strike in the United States that reduced both exports as well as customs receipts for imported goods.

The civilian council then relinquished power to Donald Reid Cabral, a conservative Dominican politician whose regime was assaulted from both the left and right by an array of political groups who "demonstrated, disseminated leaflets, negotiated, cajoled, instigated strikes, and engaged in a series of comical coups, all abortive" (Langley 1982: 253). His regime, backed by fewer and fewer Dominicans and supported mainly by his contacts with the US government, collapsed suddenly in late April, 1965, under pressure from pro-Bosch groups seeking a restoration of constitutional government and reinstatement in office for Bosch, exiled in Puerto Rico. American officials saw the complex struggle as a left–right confrontation; accordingly, US military attachés possibly encouraged the Dominican air force to strafe the National Palace (held by Bosch supporters) on April 25, an act that inflamed potential combatants on all sides (Draper 1971: 8–9). In Washington, US government officials who believed that the Bay of Pigs debacle had come from American tentativeness, urged direct military intervention. Their urgings were reinforced by the new CIA director William Raborn who explained to US congressmen that events in Hispaniola were a natural outgrowth of a "Moscow-financed, Havana-directed plot to take over the Dominican Republic" (Lowenthal 1972: 104–5).

The initial detachment of US forces (500 marines) established an International Security Zone on the western edge of Santo Domingo in order to protect the US embassy and foreign nationals. Then they linked up with the much larger US Army contingent, which arrived a few days later and then pushed through the urban area from an air-force base east of the city. American soldiers were involved in skirmishes, neutralizing rebel positions, and keeping rival Dominican factions from one another during May and June, 1965. The Dominicans held successful democratic elections in 1966, and the last American troops were withdrawn from the country in September of that year. Contrary to numerous predictions that the American withdrawal would lead to further bloodshed, the fabric of Dominican society remained intact. And Dominicans of all political persuasions now refer to the American intervention of the mid-1960s as *la revolución*, recognizing it as a pivotal event in their history (Nash 1985).

The intervention in the Dominican Republic had been the first direct involvement by US troops in the internal affairs of the region for over thirty years, but it reinforced the extraordinary American concern toward the Caribbean owing to geographical proximity and decades of thinly veiled US control that Fidel Castro had so recently shaken. In many circles the intervention was interpreted as a "costly mistake" that indirectly aided anti-progressive forces throughout the hemisphere. "For Latin America, the lesson of the Dominican intervention was that the United States, despite its stated commitment to equality and progress, would take the side of reaction when the chips were down. All the right wing had to do was to claim a Communist threat" (Robbins 1985: 3).

### The Grenada invasion

The shadow of US geopolitical domination, heretofore concentrated in the Greater Antilles and the coastal rim of Central America, extended into the eastern Caribbean after World War II as Britain granted political independence to most of its former colonies. In British Guiana, the avowed Marxist Cheddi Jagan, leader of the colony's numerically dominant "East Indian" element, was the country's Chief Minister on the eve of Guianese political independence. Fearing Communism and, again, "another Cuba" on the northeastern shoulder of South America, the US government helped the British devise a plan for proportional representation in British Guiana that led to Jagan's electoral downfall. In 1964, Forbes Burnham, Jagan's erstwhile political ally, but now a rival and leader of the country's urban black constituency, won a general election in British Guiana as the head of a political coalition. Then Burnham, whose ascendance was universally considered to be aided directly by the American CIA, became Prime Minister of an independent Guyana in May, 1966 (Jagan 1967).

Events in two other former British Caribbean colonies, recently independent Jamaica and also Trinidad and Tobago, deepened United States' government concern during the 1970s. Trinidad and Tobago, an oil-rich state and the southernmost of the Lesser Antilles, underwent large-scale strikes early in 1970 by Trinidadian oilfield, sugar estate, and service workers. The work stoppages were coordinated in April 1970, and joined by local "Black Power" groups. Trinidad and Tobago proclaimed a state of emergency on April 21, 1970, as army units sympathetic to the strikers rebelled (and were subsequently controlled by elements of the country's coast guard). A United States naval task force hovered offshore during the crisis in Trinidad (Oxaal 1982). In Jamaica, long considered by most Americans as no more than a sunny vacation spa, Prime Minister Michael Manley began to utter slogans that Americans not only did not want to hear but which also sounded ominously similar to those that had come from Cuba for over a decade. In 1974 Manley declared Jamaica a "democratic socialist" state where capitalism's legacy soon would be eradicated, a goal he attempted to achieve in part by affixing additional tariffs on Jamaican aluminum ore extracted by North American mining companies (Langley 1982: 273).

But it was neither Guyana, Trinidad nor Jamaica – the three most populous, mineral-rich, and therefore strategically most important members of the Commonwealth Caribbean – that became the focus of US concern in the former British Caribbean as the decade of the 1980s began. In 1979 a political coup in tiny Grenada led to a left-leaning government and strong ties with Cuba and Eastern Europe. Grenada's sociopolitical fabric unraveled late in 1983, and the safety of a number of US citizens on the island provided the announced justification for the invasion of the island by US forces on the morning of 25 October. The leftist government was thus eliminated, Cuban workers (and twenty-four Cuban bodies) returned home, and Grenada's experiment with applied Marxism forcibly halted.

Why, of all places, Grenada? Those few Americans who knew

Grenada existed prior to the invasion associated the island with its picture-postcard harbor surrounded by the capital town of St. George's that tumbled down to the waterfront from the green, volcanic mountains in the background. Further, Grenada's colonial legacy had been less harsh than those producing the polarized societies that were outcomes of sugar-cane monocultures elsewhere in the Commonwealth Caribbean. Grenada's economy, even in the nineteenth century, had depended almost entirely on forest crops rather than on the backbreaking regimen of the canelands: cacao had been the mainstay of the island's economy into the early twentieth century; and spices and bananas dominated thereafter. Finally, how could a country of 100,000 people and smaller (133 square miles) than most US counties pose any threat to the United States, or even to its nearest neighbors of St. Lucia, St. Vincent, and Barbados?

Attempts to address these questions underline at least two main points discussed further in Chapter 8: first, political events on every Caribbean island – though related to broad currents of external thought – are rooted strongly in local events and personalities; and second, radical changes in these tiny places often depend upon the actions and judgments of a very few and can therefore be accomplished much more quickly than in larger, "advanced" metropolitan societies.

Grenada's political transition from British colony to independent state, a transition beginning at mid-twentieth century, was associated closely with one man. Eric Gairy, a Grenadian black man of humble origins, had emigrated to Trinidad for work during World War II and to the US oil refinery on Aruba thereafter. He earned his political spurs as a strike leader on Aruba and returned to Grenada as a champion of the island's black laboring class and against the local estate owners (Singham 1968). Gairy subsequently was Chief Minister and Premier as Grenada progressed from colony to membership in the ill-fated West Indies Federation to Associated Statehood and eventually political independence in February, 1974. Gairy's flamboyant personal style was accented with an intolerance toward opposition. And his bodyguard, the infamous "Mongoose Gang," settled scores with those who disagreed or spoke out against him. Political opposition, significantly composed of large numbers among the middle and upper classes in Grenada, grew in the months preceding independence; immediately prior to independence, Grenadian police killed the father of an opposition leader, Maurice

Bishop, during civil disturbances on "Bloody Monday", January 21, 1974 (Mandle 1985: 16).

Bishop forcibly seized power from Gairy five years later. While the latter was visiting New York, Bishop and a contingent of 40–50 men armed with vintage weapons and the element of surprise took control of the police and radio stations on March 13, 1979, proclaiming victory for Bishop's party, the New Jewel Movement. Bishop declared several immediate aims that the new government intended for Grenada: social and economic improvements, freedom from local and international tyranny, and democratic elections, policies embracing the kind of social welfare objectives mirrored in "the policies of the left-wing Michael Manley government in Jamaica in 1972–80" (Lewis 1987: 28–29). But Bishop's government soon sounded a strident anti-US tone. He established early and close ties with Cuba. Further, Nicaraguan revolutionary leader Daniel Ortega was an honored guest at the first anniversary of the Grenada revolution on March 13, 1980, and Bishop was accorded a reciprocal visit to Managua the following July (Payne *et al.* 1984: 85).

During Bishop's four years in leading Grenada, he attempted to institute social reforms to improve the island's economy (Mandle 1985: 44–67). But the revolution there seems often to have floundered under the weight of its own rhetoric and its many rallies and solidarity marches (Naipaul 1984). Internal divisions in the revolutionary government were reflected in the antagonisms between Bishop (the acknowledged favorite of the Grenadian people), and the government's doctrinaire Marxist theoretician, Bernard Coard, an economist who had developed a deep, anti-imperialist sentiment stemming from his own West Indian background as a student in the United States and also England. Coard appears to have gained the upper hand in the autumn of 1983, and Bishop was confined to house arrest in St. George's. On October 19, 1983, as thousands of his supporters marched on his behalf, Bishop and several key followers were gunned down by members of the Grenadian army, and a Revolutionary Military Council assumed control, immediately implementing a 24-hour curfew (Lewis 1987: 48–66).

At dawn on October 25, 400 helicopter-borne US marines landed at Pearls Airport, on the east coast of Grenada across the island from St. George's, and shortly thereafter 800 US paratroops landed at the Point Salines airstrip near St. George's. Three hours later US President Ronald Reagan appeared on US television with Prime Minister Eugenia Charles of Dominica to announce that a "joint

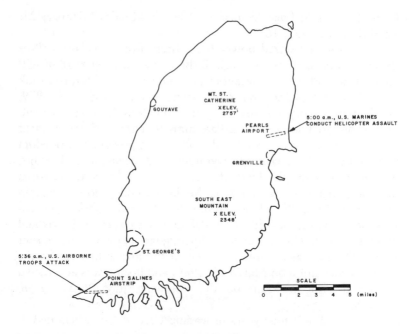

Map 4    Grenada, October 25, 1983

force" of Americans and soldiers from nearby Caribbean states had landed successfully in Grenada. Sporadic fighting continued for the next few days, although the outcome was never in doubt. Among the eventual casualties in the Grenada invasion were Cuban engineers and laborers working to extend the Point Salines airstrip.

The airstrip was under construction, according to US officials, to accommodate Soviet jets, although an improved airport for the Point Salines area to bolster tourism in Grenada had been the subject of development studies as far back as the 1950s (Payne *et al.* 1984: 31–34). A more immediate avowed justification for the invasion was the protection of 700 American medical students temporarily residing in Grenada and attending St. George's University Medical School, an offshore institution staffed by US physicians. Skeptics suggested that the American students never were in serious danger, and the protection of their welfare an ill-concealed ploy by the United States to eliminate an annoying political regime (e.g., Lewis 1987: 104–5). The great majority of the Grenadians, who had just undergone an agonizing month bordering on political chaos, welcomed the presence of the American troops (Payne *et al.* 1985: 178;

Lewis 1987: 179). International reaction was much less favorable: the United Nations General Assembly voted 108–9 to condemn the invasion. In the United States, public opinion was broadly positive, a reaction to counter the despair following the fatal bombing of over 200 US marines in Lebanon a few days earlier.

American troops left Grenada within months, although a robust US political-economic presence remained. Local elections a year after the invasion brought Herbert A. Blaize, a former Premier during Grenada's Associated Statehood, to power as Prime Minister. In December, 1986, fourteen members of the previous revolutionary faction, including Bernard Coard, were convicted of the murder of Maurice Bishop and sentenced to death, a sentence delayed by court appeals into 1990. In the first five years after the invasion, the United States had granted Grenada US $110 million in economic aid leading to improvements in agriculture and transportation. And the Point Salines airport, earlier condemned for its military capabilities was, by late 1988, a key to the island's booming tourist industry (Treaster 1988).

To those who had interpreted the US presence in the Caribbean in decline since the Dominican intervention, the popularity within the US of the Grenada episode was a reminder that most Americans still took seriously their country's paternal role over the fortunes of the region, even into the last years of the twentieth century. One interpretation saw the Grenada invasion revealing "a popular social imperialism in American life" (Lewis 1987: 191). Whatever the reasons, it was clear that constraints on local insular politics – at least in the small islands of the eastern Caribbean – still depended as much on the region's proximity to the United States as they did on either insularity or volatile local issues.

### The Caribbean Basin Initiative

The American invasion of Panama in December, 1989, in order to capture Panamanian President Manuel Noriega on drug charges, was not, strictly speaking, in the Caribbean region itself. Yet it was so close as to remind those in the Caribbean that Grenada-style gunboat diplomacy could be unleashed at any time to deter similar threats. In many ways, the formidable US presence in the Caribbean is interrelated with American economic hegemony there, an interrelationship not unlike that of the early twentieth century.

If anything, insular economies of the Caribbean are tied more closely to the United States in the 1990s than ever before. Minor changes in the huge US market (an upbeat tourist season, the invention of artificial sweeteners) can create feast or famine in tiny Caribbean states. And, early in the 1980s, the US took a diplomatic initiative to use the power of its marketplace to thwart further Grenada-style developments in the region.

The United States government's Caribbean Basin Initiative (CBI) was conceived as, very generally, an economic program intended to help Caribbean countries export goods to the United States and also to reinforce the importance of private enterprise throughout the region. Not unlike the "Alliance for Progress," a multi-million dollar program begun in 1961 and designed to spur economic growth (thereby eradicating poverty-bred discontent) throughout Latin America, the CBI has assumed that economic success, US style, will follow given sufficient economic incentive (Wiarda 1986: 141). The first years of the program have, however, produced more rhetoric than substance. As the CBI enters its second decade, the few material benefits it has bestowed upon Caribbean peoples has been far outdistanced by CBI policy pronouncements and an ever-increasing academic literature speculating as to its possible success or failure (Axline 1988a).

Although implemented by the United States, the program was inspired by a Harvard-educated Jamaican politician who emphasized, obviously for American consumption, the Caribbean's strategic position as a part of a global struggle between left and right. In October, 1979, Edward Seaga asserted that the Caribbean region faced a clear choice in economic development strategies: the free enterprise model as pioneered by Puerto Rico in close association with the United States mainland; or the socialist model exemplified in Cuba. After Seaga was elected Jamaica's Prime Minister in 1980, he visited American President Reagan who was intrigued by the idea of extending American trade and aid preferences throughout the Caribbean region, thereby tying most of the area ever closer to the United States while significantly excluding the left-leaning governments of Cuba, Nicaragua, and Grenada. In February, 1982, Reagan announced the launching of the CBI in conjunction with his personal visits to Jamaica and Barbados. The program, designed to include Central America and the insular Caribbean (with the three exceptions noted above) had three main proposed features: (1) an immediate emergency aid package, much of it for El Salvador, embroiled in a war with local leftist guerrillas; (2) a request to the US Congress to allow most Caribbean goods into the US duty-free; and (3) a strong promotion of US business throughout the region through a system of tax incentives (Erisman 1984: 13).

Caribbean reactions to the announced program were mixed: Seaga said that the CBI was "the best thing ever to have happened to the region," a pronouncement not unrelated to a projected US $50 million in aid planned for Jamaica. Maurice Bishop of Grenada termed the CBI "nothing but an insult to the region" (Polanyi-Levitt 1985: 263). Not surprisingly, the region's newspapers headlined the CBI as a possibly critical event in the Caribbean's history, and locals, at least in the Lesser Antilles, immediately began to refer to themselves self-consciously as residents of the "Caribbean Basin," a term they never before had used.

The professed enthusiasm accompanying Reagan's economic proposals soon was tempered by the realities of internal US politics. At the same time the CBI was being touted as a remedy for Caribbean economic ills, US legislators were raising domestic price supports for US sugar, thereby spurring more American production and, as a corollary, a reduction in sugar imports (Conkling 1987: 8). The American congress also killed the CBI's free-trade clauses for Caribbean leather goods, petroleum products, and textiles (Alexander 1986: 55–56). So, when the Caribbean Basin Initiative became US law in January 1984, it carried decidedly weaker provisions than Reagan had proposed. In fact, descending commodity prices were dramatically lowering, not raising, Caribbean revenue during the period in which the program was being initiated. In 1980, for example, the Dominican Republic exported US $290 million in sugar, half that in 1987; and between 1978 and 1985 Jamaican bauxite earnings fell from US $234 million to $78 million (Elliott 1987: 10).

Why, other than congressional skepticism, was the CBI not an overnight success? First, world commodity prices, in the Caribbean and elsewhere, tumbled in the early 1980s. Second, the notion that the CBI would be an extension of Puerto Rico's economic success was naive because of fundamental differences between Puerto Rico's ties with the US versus relationships between the United States and the rest of the Caribbean region: except for the US Virgin Islands, there was no free migration to the US from the other Caribbean states (as there was from Puerto Rico) for those displaced from labor-intensive jobs. Third, American investors were not anxious to commit themselves to areas such as Jamaica, Guyana, and Trinidad that seemed politically volatile (Polanyi-Levitt 1985: 272–73).

As the Caribbean Basin Initiative enters the decade of the 1990s, it is at once inappropriate to call it a success yet at the same time facile to label it completely ineffectual. If nothing else, it has called the attention of other metropolitan countries to the Caribbean region. Since the CBI began, Canada has begun a somewhat similar program (Caribcan), and Canadian business interests have increased in the region. Canada has had a notable Caribbean diplomatic and banking presence, especially in parts of the Commonwealth Caribbean, since the early twentieth century. Yet, despite recent increases in trade between Canada and the Caribbean "it is probable that future links between Canada and the Commonwealth Caribbean will be largely at the personal level of migration, education, technical assistance and tourism" (Momsen 1990: 21). The European Community developed the Lomé Agreement in 1975, an arrangement that allows duty-free entry into Europe for products from many of the Caribbean, African, and Pacific states and is renegotiated regularly (Alexander 1986: 59). And fine-tuning of the CBI by American interests still could change the region. In 1986, for example, the US Congress approved the pairing of Puerto Rican research and development laboratories with fabrication plants elsewhere in the region where labor costs are lower (Elliott 1988: 12). All of these points underline the axiom that Caribbean economic fortunes, as they always have been, are tied irrevocably to external controls, a point amplified in the following chapter.

For Americans, whose collective ignorance of the Caribbean region remains nearly limitless, the CBI represents a curious sort of discovery. They suddenly are reminded that much of the refined petroleum coming to the United States comes from Caribbean sources. Further, they are asked to look beyond the image of the Caribbean as a playground, an American "backyard," or a source only of stereotypical Cuban rhetoric, emaciated Haitian refugees, and Dominican baseball players. Yet there also is continuity with traditional assumptions because the CBI continues to advertise the Caribbean region as an appendage of American capitalism. The statement that the CBI "has offered American industrialists an attractive way to tap a plentiful supply of cheap labor conveniently waiting at the country's southern doorstep" (Conkling 1987: 3), for example, could have come in its entirety from the original labor recruiting correspondence from the Panama Canal.

American ignorance of the Caribbean is of course not reciprocated. Geographical proximity and the cumulative weight of decades of one-way economic and social domination are now heightened by the immediacy of television and radio. Throughout the islands and rimlands of the region, Caribbean peoples hear the messages of American evangelists and American politicians. Locals deal on a daily basis with visiting American tourists. Occasionally they see the international edition of the *Miami Herald* (the Caribbean's only regional newspaper), and in their own newspapers they read in detail about such locally vital matters as American congressional hearings concerning US sugar import quotas, an arrangement few Americans know of. They are dismayed, but unsurprised, that current American policy – such as that exemplified in the Caribbean Basin Initiative – is so incoherent that the United States can make grandiose promises in one breath and then take away (without knowing it) in the next. Probably most consider Americans at once well meaning yet immobilized by self-interest. As one British journalist has concluded: "Only Americans ... are ignorant of how overbearing a neighbor their country can be" (Elliott 1988: 18). From a Caribbean point of view, the unfolding of the twentieth century has made his observation an obvious understatement.

# 5

## Economic dependency

I'll tell you what, brother. If we lose our outside market, then
we're gonna have to suck an awful lotta cane.

Trinidadian sugar cane farmer, 1971

If you were to say "1992" to a resident of the eastern Caribbean's
Windward Islands (Dominica, St. Lucia, St. Vincent, Grenada), the
reaction would almost certainly have nothing to do with the 500th
anniversary of Christopher Columbus's epic voyage across the
Atlantic Ocean. Rather, the peoples of the Windwards associate
1992 with their immediate economic circumstances. That is the year
for which full economic integration of the European Community
(EC) is scheduled. Accordingly, the EC's imminent economic coalesc-
ence, as far as Windward islanders are concerned, throws open to
serious question the existing economic agreements that, since mid-
twentieth century, have allowed bananas from the Windward Caribbean to enter Britain under preferential market conditions (Thomson 1987).

On the eve of the 1992 changes for the EC, bananas from the
Windwards are not assessed import duties upon entering the British
market. They are further protected by a limitation that the United
Kingdom has imposed on so-called "dollar" bananas from Central
and South America. Although Windward islanders are by no means
wealthy, they have prospered in a relative sense because of these
special trade arrangements. During a recent period when most world
commodity prices were slumping, banana exports from Jamaica and
the Windwards increased five-fold, from US $39 million in 1977 to
US $195 million in 1986 (Elliott 1988: 10). But 1992 could end all
that. The new EC arrangements would, in theory, nullify such
special trade arrangements previously maintained by individual
European countries. British assurances to Commonwealth Caribbean banana farmers that 1992 really will not change things have so
far been only verbal (Gonsalves 1989). And Britain has only a single
voice in the EC. It is thus possible that cheap "dollar" bananas could

soon flood the European market, driving down banana prices there. Such a development would mean hardship for the tiny economies of the Windwards, and that is why so many residents of the English-speaking Caribbean view 1992 with more trepidation than antici-pated celebration.

It seems grimly appropriate that "1992" can be interpreted from the Windwards' perspective as a reminder of how vulnerable their banana economies are to external circumstances. Nor is this vulner-ability confined to either these volcanic islands of the eastern Caribbean or their relationship with the British banana market. Rather, it is the essence of Caribbean economy. Regardless of how the banana issue eventually is resolved, it is certain that similar issues will emerge elsewhere in the Caribbean in the future as they have in the past. In the early 1970s, as only one example, similar fears surrounded Britain's original entry in the EC. At that time the main issue was whether or not raw sugar from the Commonwealth Caribbean still would be allowed entry into the UK. It was, as it turned out, although happy endings in such matters for small-scale Caribbean economies have been atypical. And this regional econ-omic vulnerability, of course, is a direct historical legacy of the region's colonial experience. External economic dependency, further, is not confined to the former British Caribbean. Although Guade-loupe and Martinique are full-fledged *départements* of France, for example, French Antillean banana production is ultimately con-trolled by a Parisian company that is a subsidiary of United Brands, a US corporation. Even in Cuba, in so many ways exceptional in the Caribbean region, ongoing sugar production has rested heavily on close trade relations with the Soviet Union and Eastern Europe.

Small island size and therefore small, undiversified economies augment an historically inherited economic dependency throughout the Caribbean realm (Demas 1965; Worrell 1987). Typically, a Caribbean island state exports one or two principal agricultural or mineral commodities. In return, it receives what economists refer to as a "basket" of imported items. The diversity of items in the "basket" brought to the Caribbean is nearly always greater than the diversity of exports, an axiom that has become more evident late in the twentieth century because of heightened communications and a demand for an ever greater array of imported products by Caribbean peoples. When prices for exported commodities are high, as they were for oil exports from Trinidad in the 1970s, the outlook is bright. But when commodity prices fall, as they did for Trinidadian oil prices in the 1980s, the fall on Caribbean islands is often far and much more serious than it is elsewhere. A main reason is the Caribbean's food-deficit status, imported food commodities constituting major items in the import "baskets." During World War II peoples residing in the insular colonies of the region worried openly about possible malnourishment or starvation if trade was curtailed for great lengths of time. It is not overly dramatic to suggest, if shipping lanes were disrupted for whatever reason, similar possibilities for the future.

The combination of small size and external dependence, further, often work together to thwart local economic development in the Caribbean. Commodity price volatility is dangerous to those who would direct local infrastructures (road networks, warehouse facilities, field preparation) to accommodate a single cash crop. Similarly, an emphasis on tourism with associated financial undertakings, job training, and building ventures is precariously dependent not on fluctuating local incomes but, of course, on those same fluctuations in Europe and North America. The causal interrelationship between small island size and inevitable monocultural economies in most of the Caribbean, an issue of immense interest to economic geographers, is one that physical geographers appreciate as well. Pleasant, sunny conditions affect an entire island state; similarly, every place on the island suffers in the event of a hurricane or prolonged drought.

External economic dependency in the region affects even the small amounts of local currency that Caribbean peoples carry in their pockets. A money economy has existed in the region since slavery (Mintz 1974b: 194–206) and, in the late twentieth century, nearly all economic transactions in the Caribbean, as in metropolitan regions, are marked by the exchanges of local currency. But Caribbean money – whether pesos, gourdes, guilders, francs, or dollars – is ultimately tied to international financial markets. So local currency devaluation or inflation, which may have much more to do with financial decisions elsewhere than with local conditions, often has the insidious effect of causing Caribbean peoples to work harder (to earn more money) in order to purchase the same imported items they bought last month for less (Nietschmann 1979). In this way, as in others, Caribbean peoples are at a locational disadvantage with the developed world. They are close enough to be affected (often adversely) in nearly every way, yet sufficiently far removed from metropolitan decision-makers that their sufferings can be conveniently classified by metropolitan financiers as distant "foreign" problems.

### The agricultural dilemma: Subsistence production versus cash cropping

Tourists visiting the Caribbean apparently like being led to believe that they are sampling the local cuisine. But the morning grapefruit a tourist enjoys in Barbados is more likely a product of Florida than of the eastern Caribbean. Similarly, an offering of "poulet boucanier" at a Guadeloupe or Martinique tourist hotel often is derived from a chicken raised in France, frozen, and then shipped across the Atlantic. (And the early buccaneers probably would never recognize the cooking technique they are said to have invented!) Hotel managers throughout the region join the oft-repeated lament that foodstuffs consumed by visiting tourists are largely imported and therefore very expensive. Much of the problem is, of course, a matter of tourists' tastes. They find the mock authenticity of tourist-oriented dinner menus appealing, but they probably would not tolerate the fare consumed by Caribbean working classes. And, despite recent efforts, notably on Barbados, to introduce local foods into hotel menus, there remains throughout the region a fundamental locational incongruity between the dishes Caribbean tourists are served and the origins of the food itself.

These points are not intended to suggest that the Caribbean is without its distinctive means of food preparation; a rich variety of cooking techniques range from those accented by seasoned "sofrito" pepper sauces of the Spanish-speaking Greater Antilles to the Asian-influenced curries in the southeast (Ortiz 1967). More important in nutritional terms, however, is that Caribbean tourist and Caribbean resident alike depend heavily on imported food. The historical momentum of an introduced overpopulation during slavery is perhaps the broadest explanation for ongoing Caribbean food-deficits. The people residing in the islands never have produced all of their own food, and the ecological ruin discussed in Chapter 2 represents a physical roadblock against it.

Small-scale Caribbean food agriculture, despite its widespread presence in the region, never has competed successfully with large-scale plantation export agriculture even up to the present day. Much of the reason is that the plantation always has preempted the best land, relegating small-scale producers to marginal locales. Only in the tiniest islands where plantation agriculture has withdrawn altogether have small farmers become predominant. And even then the long arm of the international commodity market exerts control. Anthropologist Michel-Rolph Trouillot interprets the small-scale cultivators in the villages of Dominica – where bananas and also foodcrops are grown – not as a self-sufficient peasantry but as "in fact working in their own gardens for a British-based transnational corporation" (1988: 157).

Others have noted, from a variety of academic perspectives, that small-scale agriculture in the Caribbean always has suffered in its relationship with plantations. George Beckford, a Jamaican econ-omist, pointed out that, beyond land monopolization, the plantation always has had superior access to technology, external resources, and lines of credit (1972: 18–29). Geographer Frank Mills, a native of St. Kitts, reports that on his home island cultivators of small upland food gardens have suffered in more direct ways. When neighboring sugar cane plantations have applied pesticides to grow-ing cane crops or burned the canes prior to harvest (a common regional practice that reduces sharp cane leaves, rodents, and snakes from the fields), clouds of insect pests have been driven into the nearby subsistence plots (1974).

On the small Caribbean islands, food deficits are routine to the point that the monthly arrival of food vessels often are preceded by shortages or absences of food staples such as certain kinds of meat. And some of the most popular tourist destinations are tiny arid islands such as Aruba and St. Martin where virtually everything, other than sand and sunshine, are imported to serve visitors. To label such artificial situations as "food-deficit" in the sense that a large resident populace would starve without imported food is not the same as discussing the food problem in Haiti. Yet it takes little imagination to suppose what would happen to these overdeveloped tourist havens if imports were curtailed for whatever reason.

Even in the largest Caribbean states, food shortages persist owing to an unequal competition with export-oriented agriculture. In revolutionary Cuba, whose political identity may be interpreted as a reaction to foreign-dominated plantation agriculture, 80 percent of the state-controlled farmland still is devoted either to sugar cane or pasture. The Cuban government faces ongoing shortages of hard currency, and sugar for export is the major means of earning it. Not all state-run Cuban agriculture is devoted to cane. Modern large-scale rice farms in Cuba have employed advanced techniques, including seeding and fumigating by aircraft. And smaller private farms, whose presence has been alternately expanded and restricted by shifting state policy, specialize in food production. Nevertheless, foodstuffs were Cuba's most important import in the first two decades after the 1959 revolution, and there is said to be continuous and widespread grumbling among the Cuban populace concerning shortages of food staples such as coffee, beans, meat, and vegetables (Mesa-Lago 1981: 85–87; Forster 1987).

What are the actual dimensions of the food-deficit situation in the Caribbean region? Answers are varied and vague. Barry *et al.* (1984: 29) cite studies that suggest an increase, not a decrease, in imported calories into the region into the 1980s with, astoundingly, over 90 percent of food consumed being imported in some places. Part of this trend has more to do with local consumer preferences for recently introduced McDonald's hamburgers and Colonel Sanders' Kentucky Fried Chicken than it does with traditional food-deficits. My own personal research on some of the small, English-speaking islands reveals a worrisome lack of hard data as to how many calories are produced locally versus those imported.

Caribbean food deficits are manifest in the widespread malnutrition in the region, most obviously, of course, among the poor. Protein deficiencies are common, and anemia is "widespread among children under five and in pregnant and lactating women" (Barry *et al.* 1984: 29). As in the days of slavery, Caribbean malnutrition always is behind the scenes, ready to appear if economic relations with decision-makers outside the region turn sour. A case in point was Jamaica during a bout of currency inflation brought on by a balance-of-payments crisis in the late 1970s and early 1980s. Between 1978 and 1985 the incidence of Jamaican children showing signs of malnourishment increased to 41 percent, as the local cost of feeding families soared. The number of child malnutrition cases admitted to Kingston hospitals more than doubled during the same period, and school performances were startlingly lower than before (Thomas 1988: 234–36).

Yet no place in the Caribbean, or in the Western Hemisphere for that matter, compares with Haiti in terms of malnutrition. In the capital city of Port-au-Prince "it can be seen in the distended bellies of children begging for tourist coins, in the eyes of those scavenging alongside the pigs in the city dump, and in the swampy slums of La Salene" (Boswell 1989: 160–61). These are not ephemeral conditions but enduring Haitian characteristics; at the time of the American occupation over half a century ago, Haitians already were notably afflicted by interrelated malnutrition and disease. Today, as bad as malnutrition is in Port-au-Prince, it may be even worse in the Haitian countryside. Rural Haitian infants are almost universally breast-fed and, at birth, weigh the same as American infants. Yet after the first six months they are weaned on a diet of starchy soups and vegetable broths. The poor quantity and quality of their diets make reduced growth rates, disease, and apathy common after Haitian babies are weaned. By the age of five or six, if a Haitian child makes it that far, he or she has a body weight three-quarters, on average, of an American child of the same age (Lundahl 1979: 413–17).

The food-deficit dilemma is amply acknowledged by Caribbean governments, and there is no end to related rhetoric produced by regional politicians. Caribbean newspaper headlines endlessly publicize ground-breaking agricultural development programs, "Grow More Food" campaigns, and the like. But no different from the insoluble problems that exist in developed countries (the drug traffic, budget deficits), the volume of political promise usually is inversely proportional to real results in Caribbean food production. As in other Third World areas, the Caribbean food problem has spawned a robust international consulting and advice-giving industry, and every island seemingly has, in direct contrast to its food supplies, a surplus of North American or European agricultural experts whose salaries are paid by local governments or soft loans to those governments from external aid-giving sources.

But too often these foreign experts deliver short-term solutions unmatched to long-term problems. Foreign advisors introducing new varieties of carrots, potatoes, or citrus, for example, cannot reorient the internal marketing systems in Caribbean states that never have had the price supports, warehouse facilities, transportation networks, or access to bank credit to support the local production of food crops. And, as is so often the case in aid-giving programs to underdeveloped parts of the world, outside advisors are ultimately spokesmen for their own governments. The United States' Aid to International Development (USAID) agency's attitude toward Haitian agriculture is a case in point. Washington's agricultural aid to Port-au-Prince in the latter decades of the twentieth century is said to have strongly encouraged small-scale cultivators to switch from subsistence to cash-cropping, thereby intensifying their dependence on market fluctuations and external imports such as imported inorganic fertilizers and pesticides. Most damaging is that these policies seem unwittingly to have taken from the Haitians what little subsistence security they possessed (Barry *et al.* 1984: 190–91).

The disappointing results of a USAID program dealing with the Guyanese rice industry in the latter half of the twentieth century further exemplify the staggering obstacles that confront Caribbean food production. In the late 1960s Guyana routinely exported

100,000 tons of rice each year to the small islands of the eastern Caribbean. The rice was cultivated in coastal East Indian villages by small-scale producers. The relatively low quality of the Guyanese rice limited its market, however, even in Caribbean locales like Jamaica. So United States officials and the Guyanese government embarked on an extensive overhaul of the Guyanese rice industry (Richardson 1972). High-yielding rice varieties brought from Texas were viewed suspiciously by Indian rice farmers in part because they were introduced through the predominantly black government. And the economic chaos of post-independence Guyana has reduced funds necessary to maintain Guyana's complex sea-defense and water-control system, an infrastructure inherited from colonial days. Guyanese rice yields subsequently have declined rather than increasing, and by the late 1980s malnutrition was rampant in Guyana, a country whose food surpluses were sent abroad twenty years earlier (Thomas 1988: 197–98). In significant contrast, rice production from the Nickerie district of western Suriname – in essentially the same sea-level environment as eastern Guyana – has been much more successful, even in the wake of paralyzing sociopolitical developments there. A key difference between Guyanese and Surinamese rice is that the latter has a guaranteed European market for its product.

Sugar cane remains the principal cash crop of the Caribbean region as it has for nearly five hundred years. Local governments have become involved in cane production more so than in the past, and the crop is cultivated under a variety of land-tenure arrangements such as by middle-level cane farmers in Trinidad, on privately-owned estates in Barbados, and on government plantations in Guyana. The exported raw sugar (a product still refined in metropolitan countries), is subject to international prices set abroad. Cuban sugar is sent mainly to the Soviet Union and Eastern Europe under the Council for Mutual Economic Assistance (COMECON) agreement. The countries of the Commonwealth Caribbean export raw sugar to Europe under the Lomé Convention. And the International Sugar Agreement (ISA) includes major sugar trading countries including the United States, Cuba, the Dominican Republic, and the USSR. The ISA monitors free-market prices for sugar, and within the ISA the United States maintains its own sugar quota – with prices usually above the nominal world price – unilaterally deciding how much sugar other countries may import to the US market (Barry *et al.* 1984: 23).

It is axiomatic, as Trinidadian cane farmers and all other Caribbean sugar producers know, that a guaranteed market for their product is vital. The point was reaffirmed eloquently for most of the region when the United States limited imports in 1981. The curtailment occurred during a two-year period when the world sugar price was plummeting from US 45 cents/pound (1980) to 8 cents/pound (1982). The results were devastating for, among other places, St. Kitts-Nevis where half the workforce in sugar production was laid off almost immediately. In the Dominican Republic, the downturn in sugar production could have long-term geopolitical implications. In 1981, the Dominican Republic's sugar quota to the US was 493,000 tons, a figure cut to 123,000 tons by 1988. In that year, the Dominican Republic signed a long-term contract to supply raw cane sugar to the Soviet Union (Elliott 1988: 12–13).

Cuba's special relationship with the Soviet Union does not protect her from the vagaries of world sugar prices. The USSR is the world's leading sugar producer (from sugar beets) and exports sugar of her own. So some of what Cuba sends to the USSR is reexported on the open world market. Although Cuba exports most sugar to the eastern bloc, she also maintains trading relationships with other countries, notably Japan and Spain. In 1974, a year of record prices on the world sugar market, Cuba realized triple the income from the same volume of exports as in 1971. Cuba is the world's leading sugar exporter and the fourth leading overall producer (after the USSR, Brazil, and the United States), although secrecy regarding Cuban sugar data poses difficulties for analyzing sugar trends there.

Bananas are the second leading agricultural export of the region. The shipment and marketing of Caribbean bananas is handled by transnational corporations and, as discussed above, a protected market for the crop is vital and subject to imminent change. Since mid-twentieth century, Geest Industries Ltd of Britain has helped to develop the Windwards into the leading banana production zone in the region. The United Brands company of the United States purchases bananas in Jamaica, Belize, and Suriname through its British subsidiary, Fyffes, as well as from the French Antilles (Barry *et al.* 1984: 39–43).

The banana industry of the Lesser Antilles historically has been romanticized through tourist-oriented calypsoes as a local livelihood involving the pre-dawn loading of sailing schooners by sturdy peasant farmers. More recently it has been modernized by market demand and sophisticated transportation techniques directed by external forces. The Geest company buys the entire banana crop from Dominica, St. Lucia, St. Vincent, and Grenada in agreement with the regional Windward Islands Banana Growers Association (WINBAN) which represents some 16,000 to 20,000 small-scale banana producers in the four islands. Green bananas are loaded onto weekly and fortnightly Geest refrigerator ships where they ripen enroute to Britain. Concerns about local handling in the islands has led to the "field-packing" technique whereby farmers pack the fruit into cardboard boxes in their own plots before hauling the produce to port. Geest officials have declared field-packing a success, although many banana farmers complain that it really represents unpaid work on their part – on top of the risk of cultivation – for the 10 percent they receive of the bananas' eventual sale price. Not incidentally, market demand also has led to the universal use of inorganic fertilizers and imported fungicide sprays in Windward banana production, innovations creating further reliance on the outside world not only for markets but also for agricultural inputs (Thomas 1987: 45–58).

Other Caribbean agricultural exports are subject to a variety of external constraints. Small-scale Haitian coffee production, the principal source of cash in the countryside, is controlled by local speculators and sustains a tax imposed by the Haitian government even before it enters the international coffee market (Girault 1985). Limited amounts of coffee also are produced in Jamaica and the Dominican Republic. Besides bananas, Geest also purchases citrus from Dominica and mangoes and rootcrops from St. Lucia and St. Vincent. Grenada sends its spices to British and US markets. Cacao still is produced in central Trinidad, although it cannot possibly compete with the enormous output of cacao from West Africa. All of these local specialty crops are marked by limited and varying outputs and, most of all, their production is inhibited by the lack of guaranteed markets outside the Caribbean region.

## Minerals

In the early 1970s, when American motorists queued in pre-dawn service station lines and wallowed in self-pity over the shortage of fuel they always had taken for granted, a related but completely different atmosphere prevailed in the Netherlands Antillean islands of the southern Caribbean. On Aruba the Lago (Exxon) oil refinery and on Curaçao the Royal Dutch Shell oil refinery prospered as world oil prices soared. Both islands also realized beneficial side effects. Free-spending Venezuelan tourists, their incomes swelled by that country's high-priced oil resources, flocked to Aruba. And a heightened volume of sea traffic into Curaçao meant more business for the island's dry-dock and ship-repair facilities.

Neither Aruba nor Curaçao has oil of its own nor does either island have other notable agricultural or mineral resources. But colonial location strategies at the start of the twentieth century led to the siting of oil refineries in each place. Earliest Dutch traders established salt works on these desert islands off Venezuela's coast and, as noted in Chapter 3, Curaçao became the center of an intense regional trade thereafter. Then, when American and European oil interests discovered and began to develop the massive petroleum reserves of Venezuela's Lake Maracaibo early in the twentieth century, the expensive refineries for crude oil were located not in politically volatile Venezuela but in the tiny, safe, Dutch islands offshore, on Aruba in 1918 and Curaçao in 1929. Venezuela eventually came to refine most of the Maracaibo oil, although into the 1980s Aruba and Curaçao remained such important refining centers that "many people . . . found it impossible to believe that there might be an Aruba without Exxon or a Curaçao without Shell" (Treaster 1985).

But the overused "boom and bust" slogan that is so applicable to Caribbean agriculture is perhaps even more appropriate for the region's scattered mineral industries. The high oil prices that led to economic boom in Aruba and Curaçao also led to increased refining in the Middle East and refinery competition in the Caribbean itself such as the expansion of the enormous Amerada Hess refinery at St. Croix in the US Virgin Islands. When oil prices plunged in the early 1980s, the unthinkable occurred in Aruba and nearly so on Curaçao. The Aruba refinery closed early in 1985 because of financial losses leading immediately to massive layoffs, the reduction of government salaries, and predictions of unemployment rates of up to 40 percent (Diederich 1985). In Curaçao, the government bought two-thirds of Shell's "money-losing" refinery (which had repatriated profits to Europe for over half a century) so as to lease the operation to the national petroleum company of Venezuela, thereby saving nearly 2,000 local jobs.

Most oil traditionally refined in Aruba and Curaçao was destined for the United States, as is much of the petroleum processed at the other Caribbean oil refineries which vary in size and capacity throughout the region (Barry *et al.* 1984: 96-99). Estimates vary, but something over half of the refined petroleum imported to the US – including oil from African and Middle Eastern sources – passes through Caribbean refineries (Payne *et al.* 1984: 44). The irony is that much of the Caribbean itself is deficient in marketable mineral resources and therefore energy poor. So it is common for people on small islands – where electricity is limited, expensive, and unpredictable – routinely to observe passing oil tankers heading north to energy-rich North America. Small island size and limited demand for energy increases its overhead cost throughout the region, conditions that inspire continuous "small is beautiful" discussions exploring the possibility of wind and solar power for Caribbean islands. In the meantime, the real-world volatility of world petroleum prices hits hard where all fuel is imported. In January, 1985, there were riots in Kingston, Jamaica, because of government-decreed increases in fuel prices and also similar disturbances in the Dominican Republic (Bonnet and Calderon-Cruz 1985).

Trinidad and Tobago is the only Caribbean country with an important indigenous petroleum industry. Local asphalt was used for all-weather highways in Trinidad as far back as the early nineteenth century, and marketable subterranean oil reserves were located there early in the twentieth century. They were later augmented by offshore drilling, the installation of local oil refineries, and the discovery of new natural gas reserves in the 1970s, so that by the 1980s petroleum and petroleum-related products accounted for more than 80 percent of the country's exports. As important as the production of local petroleum was, even more important was the Trinidadian oil-refining industry which processed crude oil from Africa and the Middle East and was responsible for some of the Persian Gulf's oil "leakage" into the United States during the oil embargo of the 1970s.

The sudden jump in oil prices at that time created economic euphoria ("capitalism gone mad" according to one calypsonian) in Trinidad, but the rapid descent back down the price ladder a decade later was all the harder after the boomlet of prosperity. Between 1974 and 1983 Trinidad and Tobago received an estimated US $17 billion in oil revenues. The massive infusion of capital allowed the government to extend an import substitution policy begun immediately after political independence in 1962; a collaborative effort between the Trinidad and Tobago government and several transnational corporations in the 1970s led to an enormous industrial complex at Point Lisas in southern Trinidad. It featured refineries, a chemical fertilizer plant, and an iron and steel works, all giving rise to smaller manufacturing industries producing cement, household appliances, television sets, and automobiles. Public expenditures on highways, medical facilities, and the construction of a new sports complex paralleled the industrial boom. Then the sudden fall in oil prices early in the 1980s led to severe cuts in social expenditures, job retrenchment, and price inflation for imports that earlier had been subsidized. The downturn in local agriculture, owing to government mismanagement and a lack of farming enthusiasm during the oil boom, reduced the buffer that might have cushioned the jolt of low oil prices. In Trinidad, the exuberance of the 1970s was replaced harshly and suddenly by a malaise in the 1980s with the burdens of economic depression borne disproportionately by the poor (Thomas 1988: 279–99).

The production of bauxite (aluminum ore) and its dehydrated variant, alumina, in the Guianas and the Greater Antilles have traditionally been important not only in the Caribbean region, but in the overall world mineral market as well. In 1914 the Demarara Bauxite Company was formed by American and Canadian mining interests, and they began operations sixty miles up the Demarara River from Georgetown in British Guiana. In 1928 the neighboring Dutch colony of Suriname signed an agreement with the Aluminum Company of America (ALCOA) for the mining of similarly situated upriver bauxite deposits, and a Dutch mining company staked out bauxite claims in Suriname thereafter. Bauxite from the Guianas was vital to the American World War II effort, and Suriname emerged from the war as the world's leading producer (Chin and Buddingh' 1987: 120). Exploration geologists found bauxite in Jamaica's interior in 1942, and during the next three decades North American companies moved in. Markets for Caribbean bauxite boomed in the 1950s and 1960s and development proceeded apace, although a downturn in prices became obvious in 1967 (Beckford 1987: 43–44). At that time all Caribbean bauxite production was controlled by seven mining companies, six American and one Canadian (Thomas 1988: 111). Most important, the final product – metal aluminum ingots – was being produced not in Caribbean locales but in North America where electricity generated by oil and gas in Texas and Louisiana and by running water in southern Canada represented the cheap energy necessary to separate electrolytically aluminum from its oxide. As in the production of Caribbean sugar, the value-added final product was thus created in the metropoles.

In 1974, Jamaica's Prime Minister, Michael Manley, imposed a levy on transnational bauxite mining companies operating there, a tax based not on the ore itself but on the selling price for aluminum ingots. He also was instrumental in establishing the International

Bauxite Association, a producers' organization formed in the same spirit, though eventually without the power, of OPEC. Jamaican tax revenues from bauxite companies increased from US $30 million annually from 1970–73 to US $196 million annually in 1978 and 1979. But US corporate reaction was hostile and effective. Jamaican production was curtailed, and intergovernmental pressure from North America is said to have influenced the political demise of the Manley regime in favor of pro-US Seaga forces in 1980 (Thomas 1988: 112–13, 214–15).

Bauxite's fortunes sagged in the Guianas as well. In Guyana (erstwhile British Guiana), Prime Minister Forbes Burnham nationalized local holdings of the Aluminum Company of Canada in 1971 as part of a national economic policy dubbed "Cooperative Socialism." He nationalized the Reynolds Bauxite Company (USA) mines on the Berbice River three years later. Subsequent government mismanagement and falling world prices struck the Guyanese bauxite mines so that the workforce was cut by one-third in 1983, the industry bedeviled by a series of work stoppages, and national output in the 1980s was roughly half what it was during the 1960s (Singh 1988: 108–11). In the mid-1980s Suriname's once-flourishing bauxite production had ceased entirely because of political chaos and associated civil unrest. By the late 1980s Caribbean bauxite – which once accounted for two-thirds of the world's production – comprised no more than one-sixth the world total (Thomas 1988: 105, 111).

The story of twentieth-century bauxite production in the Guianas is in many ways the story of the Caribbean economy. Both Guyana and Suriname have immense potential for hydroelectricity in the mountainous backlands, and for years development reports bristled with predictions of projected industrial integration schemes in which the combination of realized hydroelectricity and local bauxite would lead to aluminum production and associated industrial prosperity. But profits from Guianese bauxite have accrued to European and North American corporations rather than being reinvested in the development of local hydroelectricity projects. International mining companies have, further, reacted predictably to local political problems by moving elsewhere, leaving both Guyana and Suriname with an inheritance of scarred landscapes and antiquated mining machinery. In Jamaica, the ecological legacy of bauxite mining is similarly stark. All but one of the transnational mining companies was gone by 1987, leaving a series of red mud lakes in their wake and the Jamaican government as the only bauxite producer. Most of the companies had complied with the Jamaican legislation calling for the restoration of the land after their operations ceased; yet 82 percent of Jamaican highland farmers, when surveyed, considered the land in worse condition than before, even after restoration (Beckford 1987: 33–35).

Except for tiny petroleum fields and local quarrying and landfill operations, the Lesser Antilles besides Trinidad are devoid of notable mineral production. Greater Antillean bauxite deposits on Haiti and the Dominican Republic were exploited until 1982, when the foreign-controlled mines were closed. The latter country produces nickel ore as does Cuba, mainly along the north coast of Oriente province. Cuba ranks fourth in world nickel ore production after Canada, the USSR, and New Caledonia. The ore-processing plant at Nicaro was opened in 1943 by the Freeport Sulphur Company of the United States. The Cuban ore was exported and then processed into metallic nickel in the US mainly during World War II and the Korean War for the fabrication of heat-resistant metal alloys used for weaponry. Since the Cuban revolution, most of the country's nickel ore goes to the USSR, although some is sold to Spain and Italy (Moran 1987).

### Industry

As discussed in Chapter 3, the Caribbean plantation traditionally has carried with it a quasi-industrial atmosphere featuring a rural proletarian workforce. And economic linkages with the outside world, such as with international marketing systems for agricultural staples, have sustained continuity with earlier days. The region's peoples as one result are thus not rustic or "backward" but modern in many ways. It would therefore appear that, without the obstacles that tradition or ignorance pose, the Caribbean is socially well-suited for modernization or development. Yet, except for export-oriented enclaves such as in southern Trinidad, the region lacks indigenous industry. In most places, small factories are confined to the minor fabrication or processing serving tiny local populations – breweries, cement plants, rudimentary crop-processing concerns, and perhaps a grain mill at dockside. Small local garment shops produce cheap clothing, but part of being a resident of the Caribbean is the aspiration to wear clothing imported from abroad because it is invariably considered of higher quality than that produced at home.

Why has the Caribbean lagged so far behind industrially? Much of the answer may be found in historical momentum and because the region continues to be controlled economically by outsiders. Caribbean business, manufacturing, and retail concerns are metropolitan appendages, truncated local enterprises where entrepreneurial success is reduced by foreign control. This truism is exemplified in many ways throughout the region. Jamaican businessmen, no different from Jamaicans who have become notably successful in New York, are said to lack "opportunity structures" in their own country (Berger 1984: 9). Also in Jamaica, a recent survey of the famous "higglers" or market-women – whose legendary distribution of garden-plot surpluses recently has extended to the marketing of manufactured imports – suggest that their aspirations are to be free of this work because of its constraints and socioeconomic limitations (Le Franc 1988).

Puerto Rico appears to be the obvious exception to the contention that Caribbean industry is structurally stunted. By the late 1980s Puerto Rico was, as measured by conventional indexes, as industrialized as countries in southern Europe. Nearly half of Puerto Rico's gross national product now comes from manufacturing, and the goods produced range from heavy machinery to mechanical heart valves. A robust tourist industry augments local manufacturing. Evening traffic jams in the San Juan area clog controlled-access highways that serve large pharmaceutical plants, no different from scenes in the hinterlands of the largest US cities. Even rural Puerto Ricans now obtain the bulk of their food at air-conditioned supermarkets. Agriculture, the mainstay of the Puerto Rican economy until the 1940s, accounts for less than 5 percent of the island's economic output half a century later.

The key to Puerto Rico's economic boom, of course, has been its special relationship with the United States. The Estado Libre Asociado de Puerto Rico (freely translated as "Commonwealth") experienced a major economic transformation, the so-called "Operation Bootstrap", in the 1940s and 1950s. Lured by tax incentives, United States industries established small factories throughout the island's countryside. Accordingly, major changes in Puerto Rico's social geography saw working men abandoning their small farm plots and their jobs on agricultural estates for nearby factory positions. Women, formerly working for pittances in their homes – often by doing needlework – took similar manufacturing jobs (Silvestrini 1989: 160). The small-scale Puerto Rican manufacturing plant of the 1950s subsequently has given way to the "superfactory" of the 1980s, a complex usually located in greater San Juan and similar in scale and technology to the giant sugar central of earlier days (Weisskoff 1985: 54).

Profit maximization by US corporations is at the heart of Puerto Rico's industrial success, profits increased by US tax laws. An estimated 100,000 new jobs in Puerto Rico between 1976 and 1985 were attributed to tax incentives explicated in section 936 of the US Internal Revenue Code. Put simply, 936 exempts US corporations from taxes derived from their Puerto Rican operations and allows the sheltering of funds for Puerto Rican industrial development (Elliott 1988: 12). Opponents of the 936 Law (that makes Puerto Rico such an attractive locale for US business investment) point to the accounting practice of "transfer pricing" which increases corporate profits; large US business concerns, operating under Section 936, can sell raw materials cheaply to their Puerto Rican subsidiaries and pay the same subsidiaries inflated prices for the final products, thereby increasing (tax-free) Puerto Rican profit margins (Barry *et al.* 1984: 249).

The Caribbean Basin Initiative, discussed in Chapter 4, is not the first case in which the representatives of nearby Caribbean states have considered the Puerto Rican case a model of success that might invigorate their own economies. The Nobel laureate and St. Lucian economist W. Arthur Lewis in 1951 envisioned similar development for the small islands of the Commonwealth Caribbean. An international development company centered in, perhaps, London or New York, according to Lewis, might unleash the economic talents of the peoples of the English-speaking Caribbean, similar to that which was occurring in Puerto Rico (Weisskoff 1985: 89–90). Eric Williams, Prime Minister of Trinidad and Tobago from its independence in 1962 until his death in 1981, also spoke often and openly of transforming his own country into "another Puerto Rico" by inviting foreign capital (Langley 1982: 272–73).

Yet, behind the seeming prosperity underlying San Juan traffic jams and the glitter of the nightclub shows in the swank hotels of San Juan's Condado Beach district, an emptiness in the Puerto Rican development experience might warn neighboring Caribbean states that are considering emulation of the Puerto Rican "model". The comparative prosperity of Puerto Rico has not been derived from true indigenous economic development. Rather, it has relied precariously on a dependency relationship with the United States. More specifically, manufacturing in Puerto Rico has not led to a proliferation of minor industries that provide raw materials and parts. Usually, it is a matter of parts and technology from the mainland aggregated in Puerto Rico for assembly and reexport to take advantage of existing tax laws.

Food production and general Puerto Rican agriculture have suffered most of all. Foodstamps issued by the US government to bolster low-income diets first came to Puerto Rico in September, 1974. Six years later nearly 60 percent of the populace was receiving them. Foodstamps and Puerto Rican unemployment (23 percent in 1983) now reinforce one another; day laborers often miss work or fail to report income so as to qualify for the food subsidies. Foodstamps, moreover, have all but eliminated Puerto Rican subsistence agriculture. Local diets are no longer satisfied with tropical food staples but with frozen and packaged foods imported through large US supermarket chains that have, in effect, undermined Puerto Rican agricultural development. The overall Puerto Rican economy recently has been interpreted as a "geographical laundry for corporate profits to avoid federal taxes", with its people living "peaceably on welfare or not so peaceably in the ghettos of America's industrial cities" (Weisskoff 1985: 89).

Elsewhere in the Caribbean, small-scale "industries" – from the sorting of US grocery-store coupons to the hand assembly of television parts – take advantage of low wages and proximity to the US market. European and North American companies and local subcontractors, who produce parts and components for the transnational corporations, employ low-paid workers who toil without contracts and under sweatshop conditions (Barry *et al.* 1984: 59–61). Under the new CBI arrangements, the pool of cheap Caribbean labor has become a target for possible Japanese investment to avoid textile quotas imposed by the United States on Asian imports, developments watched closely by American legislators (Steele 1988).

The worst regional conditions for small-scale factory workers are in Haiti, the place where – as every American schoolboy knows – most US baseballs are manufactured. In the 1980s some 60,000 men, women, and children worked in 200 small assembly plants in Port-au-Prince producing electronic products, garments, and sporting goods, using materials imported from the United States. Transnationals and subcontractors alike kept wages incredibly low (an average US $3.12/day in 1984), in part by employing trainees and then laying them off when their "apprenticeships" were completed (Dupuy 1989: 175–79).

Low Haitian wages and peaceful business surroundings are, it must be emphasized, requirements for the location of these factories, not simply local cultural traits that then appeal to foreign entrepreneurs. Jamaican workers recently learned, for example, what happens when conditions are less than optimal for footloose foreign businesses. The Jamaican government in 1960 provided factory space and tax holidays for foreign assembly firms which then took advantage of low Jamaican wages for more than a decade. When Jamaica began to ask for higher wages and more benefits from the foreign companies in the mid-1970s, many of the small factories closed, transferring their operations to more amenable Caribbean locales (Thomas 1988: 84–85).

## Tourism

The jet airplane descends before passengers can see the island of Antigua, and landscape features are recognizable only on the final approach to Coolidge Field. The asphalt runway, constructed during World War II, has since been extended greatly to accommodate the largest passenger jets. Now the airport is a vital element in Antigua's booming tourist industry; over 17,000 tourists arrived in Antigua (population 80,000) in March, 1986, alone – 5,000 more than came in all of 1958 (Elliott 1988: 15; Henry 1984: 123). The passenger who arrives in Antigua either from New York's JFK airport after a four-hour flight or from London's Heathrow airport (in 7½ hours) initially may be disappointed in the parched brown landscape. But Antigua's reputed 365 beaches (a figure suspiciously identical to the number of rivers claimed by Dominica), relentless sunshine, and carefree atmosphere are the requisites of a vacation paradise. From June through October, 1989, as advertised in *The New York Times Magazine*, a North American family of four could enjoy the "highly attractive summer rates" of US $340/day at Antigua's beachside St. James's Club (meals not included). A six-day stay for this hypothetical family would therefore cost about US $2000, a figure, incidentally, equal to the estimated per capita annual income of the average resident of the two-island state of Antigua and Barbuda.

Antigua's tourist industry benefited accidentally from the Cuban Revolution of 1959 which redirected American tourists elsewhere. And by the 1960s it was obvious that tourism in Antigua was becoming an economic staple for the island. The "Hotels Aid Ordinance" already had been passed by the island's colonial government in 1952 to reduce import fees for hotel construction materials, and a decade later expensive airport, road, and harbor improvements were undertaken to improve Antigua's image. Despite tourism's growth in Antigua, not all of the results have been beneficial. High-paying tourist-related jobs have almost always gone to foreigners, with Antiguans themselves serving as waiters and chambermaids; tourism has not forged links with other sectors (such as agriculture) of the economy; and substantial profit "leakages" have flowed from the island to foreign investors and promoters (Henry 1984: 121–27).

Laws similar to the Antiguan statute limiting duty on hotel imports as well as tax-holiday regulations are found throughout the Caribbean because of tourism's paramount economic importance. Roughly 11 million people, including cruise passengers and "stop-overs", visit the region each year. Tourism is the only activity of note in some smaller islands and, in good years, the leading earner of foreign exchange even in Jamaica (Elliott 1988: 15). In the 1950s when Caribbean vacations were – except in Cuba – exotic and adventuresome sojourns involving small, stucco, family-run hotels, travelers from Britain, France, and the Netherlands frequented resorts in their own Caribbean colonies in high percentages. In the subsequent decades of glass and steel highrise hotels, each featuring hundreds of rattling air-conditioning units, the Caribbean tourist industry has come to be dominated by Americans; in 1985 two-thirds of all tourists visiting the region came from the United States (Thomas 1988: 151).

The "development" of some tiny Caribbean islands to accommodate large-scale tourism has produced startling changes in the latter part of the twentieth century. For example, the Dutch half of tiny St. Maarten – an island shared with France – is only sixteen square miles with an indigenous human population of about 13,000. Occasional visitors to St. Maarten since 1970 have seen it transformed from a quiet islet of small hotels to a mammoth beach resort where New York accents, traffic jams, and sanitation problems prevail. In 1980, St. Maarten received 179,000 tourist visitors, in 1986 an astounding 439,000 (Elliott 1988: 15). It hardly needs emphasis that the demand for fresh water, paved roads, flush toilets, and garbage disposal strain the carrying capacities of St. Maarten and other tiny places beyond their abilities to accommodate the attendant ecological stress. Few islands have escaped water pollution, beach erosion, and the proliferation of junk heaps, which together represent the apparent environmental costs of Caribbean tourism.

The nature of large-scale tourism helps explain why it is subject to a growing and even more intense external control than are other sectors of the Caribbean economy. The tourists themselves who reside in metropolitan areas choose or are guided to Caribbean destinations from occasional personal experience but more often from advertisements and existing airline schedules. A given island's tourist income is roughly proportional to the number of tourist arrivals, although the number must be kept manageable so as to avoid over-crowding which will discourage visitors from returning or, worse, antagonize travel agents. The job of matching potential tourists with particular islands usually is not a Caribbean enterprise but handled in the computerized reservation systems maintained by such companies as Air France, Air Canada, or Sheraton Hotels. Accordingly, package tours and air travel/hotel combination holidays usually involve pre-payments not to Caribbean hotel owners but to, for example, Thomson Travel in Great Britain or American Airlines in the USA. Further, transnational hotel corporations extract much of the profit from Caribbean tourism without necessarily owning the hotels themselves. Caribbean governments and hotel owners often purchase licenses from corporations like Holiday Inns of the United States who train service personnel and also give advice as to what this year's tourists will expect from their local hosts. Expensive television advertisements – crafted by North American and European agencies and designed to capture just the right combination of warmth, pleasure, and personal safety – further extract tourist profits from the Caribbean to the benefit of the developed countries of the North Atlantic (Barry *et al.* 1984: 76–81).

The advertisements are crucial because they must portray friendly, sunny destinations while also targeting various segments of metropolitan populations, and miscalculations can be costly. Caribbean vacations must not come across as too expensive to middle-class North Americans lest they choose instead the comparatively cheaper resorts of the Mexican Gulf Coast. At the same time, too many middle-class or blue-collar Americans or Europeans flying cheap package tours to a given island can repel wealthy, class-conscious visitors. Instead of paying US $1000/day to stay at an exclusive hotel in Barbados, wealthy families from London or New York might choose, for example, Mauritius in the Indian Ocean which advertises itself as "not for everyone."

Subtleties in advertised images projected abroad, swings up or down in metropolitan stock markets, or particularly cold or warm winter weather in North America or Europe can create success or failure in a given island's tourist season because small-scale insularity and tourist volatility go hand in hand. A passing hurricane, a series of volcanic earth tremors, or a well-publicized rape or murder – if the victim is a white tourist – can drive tourists elsewhere with disastrous local results. The grisly slaying of eight people by five young black men at the posh Fountain Valley golf club in St. Croix in September, 1973, helped lead to a flurry of canceled airline reservations, and hotel occupancy was down by one-third in St. Croix during the ensuing winter season (*Dollar impact* 1973). But the volatility factor can also produce good fortune in a Caribbean tourist economy. Barbados was suffering a slow, lifeless tourist season in 1981–82, presumably in response to economic recession in the United States. Then US President Ronald Reagan's visit in April filled Bridgetown area hotels for two weeks with American news personnel, security agents, US State Department officials, and assorted hangers-on so that, according to some Barbadian hoteliers, the island's 1981–82 tourist season was transformed by Reagan's visit alone into a marginal success.

Billboards throughout the region remind (black) local residents to put on happy smiles for (white) tourists. In their quest to get away from it all for a week in the winter, white Americans want no experiences with black hostility which they feel they already know from their own country. So groups of tourists can be typically loud and offensive while expecting deferential servility from their "hosts." Caribbean governments, with an eye on tourist profits, reinforce these expectations. It is perhaps needless to point out that this economically imposed servility is galling in light of the obvious (at least to Caribbean peoples) inequities mirrored in the juxtaposition of luxurious tourist hotels that are the domain of white tourists and wooden shacks occupied by local peoples. Guyanese economist Clive Thomas asserts that the development of Caribbean tourism, accomplished at enormous monetary and social costs, actually contributes little to change "the widespread poverty and powerlessness of the West Indian people" (1988: 167).

### The narcotics traffic

Because relatively poor Caribbean peoples are so very aware of their proximity to the conspicuous riches of North America, it is not surprising that they maintain a kind of "treasure chest" mentality. A US $100 tip by an inebriated tourist or, even more doubtful, a dropped wallet or pocketbook brimming with money, are vague hopes nurtured by some local men and women locked into dead-end hotel service jobs. A more vivid example comes from the tiny island of Carriacou in the Grenada Grenadines: during World War II a floating harbor mine is said to have ridden the easterly swell across the Atlantic from North Africa eventually to beach at Carriacou; curious island residents, convincing themselves that the metal globe contained money from America, are said to have cracked open and thereby detonated the device, killing several bystanders. Late in the twentieth century, hopes of lost treasure bound for North America have rekindled similar, but much more tangible, hopes among residents of the Bahamas. Because of the massive drug traffic between the many Bahamian islands and cays and southern Florida, a traffic featuring clandestine reloadings of illicit cargoes and high-speed boat chases, residents of the Bahamas "commonly hope that some lost duffled bag (filled with drugs) will wash ashore and make them rich" (White 1989: 39–40).

The illegal drug smuggling into the United States, arguably the most pressing American social problem late in the twentieth century, is vitally linked to the Caribbean region. Relatively little of the material is produced in the Caribbean itself, but the route from the source area of cocaine in northwestern South America and Miami is a straight line bisecting the Caribbean's heart. Cocaine, a white or colorless narcotic alkaloid, is derived from the leaves of several species of coca shrubs of the genus *Erythroxylon* grown in low to intermediate altitudes in Andean South America. Andean peoples have chewed coca leaves since at least 1500 BC, its stimulating effects said to compensate for harsh, oxygen-poor Andean environments. But the contrast between the physical stimulation derived from chewing a coca leaf and the rush of clairvoyance from ingesting cocaine has been likened to the difference between travel by donkey and jet aircraft. In any case, the main area for coca cultivation is the eastern slopes of the Andes in Peru and Bolivia, although Colombia dominates the cocaine production and smuggling. And the astonishing sums of American money accruing to Colombian sources as payment for smuggled cocaine has helped produce rampant crime and corruption there on a massive scale (Gugliotta and Leen 1989).

Along the route from Colombia to Miami, cocaine is transported illegally (according to a US law passed in 1914) in every way imaginable. This smuggling often involves transshipments into the United States from, increasingly, Haiti and the Dominican Republic, although the principal zone for transshipment during the 1980s has been the Bahamas. The presence of cocaine in the Bahamas is so pervasive that there is "a crack house on every corner" and, metaphorically speaking, wooden shanties in Nassau without electricity but with Mercedes-Benzes parked outside and "twenty kilos of cocaine under the bed" (Eddy *et al.* 1988: 133).

The logistical necessities of offloading, repacking, and onloading smuggled drugs plus the need for refueling water and aircraft in the Bahamas – activities all associated with the transfer of enormous sums of money – have led to payoffs and corruption at every level of society. Officials throughout the Bahamian government, most notably the Prime Minister, Sir Lynden Oscar Pyndling, have been accused of routinely accepting bribes (Eddy *et al.* 1988). Police in the Bahamas are said to be so involved in aiding cocaine smugglers that to characterize their behavior as "looking the other way" would be, strictly speaking, inappropriate because usually they are on the scene to help things along and, on occasion, assist with loading operations! Yet drug-related corruption in high places is not necessarily a Bahamian monopoly. Three ministers from the government of the Turks and Caicos islands were arrested on cocaine charges in Miami in 1985. Further, in the 1980s there have been drug-related scandals involving both high political officials and police officers in the Cayman Islands, Jamaica, Trinidad and Tobago, and Barbados (Thomas 1988: 170). The most celebrated recent case of drug-related corruption in the region has occurred in Cuba. In mid-1989, Arnaldo Ochoa Sánchez, former commander of Cuban Forces in Angola, was arrested, tried, and executed owing to his association with drug smuggling. Videotapes of portions of Ochoa's trial appeared on Cuban television, and some interpret his execution as evidence of a hardline policy decreed by Fidel Castro against any disagreement or deviation from his regime (Preston 1989).

Besides the bribery and corruption in parts of the region that have together provided an illicit support system for drug smuggling, technically legal institutions also have profited and proliferated from the trade. Offshore banking in the Caribbean facilitates the laundering of drug-related money; a notice said to hang in the lobby of a Cayman Islands bank informing patrons that "We take no pictures" assures camera-shy couriers that their transactions will be handled quietly (Barry *et al.* 1984: 132). Similarly, banks in the Netherlands Antilles process bales of US currency with few questions asked (Frank 1986). These points are not to suggest that every Caribbean banker will automatically produce a cashier's check in return for a valise packed with cash. Yet the sudden presence of so much drug-related money in the region doubtless is the basis of interest and envy to the representatives of the foreign banks that have dominated the region's finances for decades.

The branches of foreign banks – mainly American, Canadian, and British – control the money flow in the Caribbean and have exerted considerable economic influence for years in the region. In some ways, local bank managers and local banks – whose continuities are not dependent on periodic elections – exert more power than do ephemeral politicians. The banks work closely, of course, with consultants and financiers representing agricultural interests, mining ventures, and footloose manufacturing concerns, thereby reinforcing export-oriented activities. Their ready access to the global capital market and economic intelligence reports predicting international swings in commodity prices are further advantages possessed by foreign banking concerns. Foreign banks in the Caribbean also exert a powerful social presence, from the ubiquitous radio advertisement jingles to having the most formidable buildings in town.

Caribbean academics and politicians occasionally have targeted the branches of foreign banks as uninterested in financing truly local business ventures, preferring less risky, externally oriented loans. But a noteworthy development from the Jamaican drug trade suggests that the capital derived recently from the narcotics industry may be producing beneficial change, at least for some black Jamaican entrepreneurs. Black Jamaican businessmen for years have experienced difficulty in securing loans from local branches of foreign banks. But in the late 1980s there are reported cases of a few black Jamaican businessmen who have used money from the local ganja (marijuana) trade to finance legitimate economic enterprises, thereby using drug money to circumvent the local financial domination traditionally exerted by transnational banks (Stone 1988: 33).

Romanticized by reggae music and a sacrament in the island's Rastafarian religious sect, the use of ganja is widespread though illegal in Jamaica. And although the cash income from ganja exported to North America is dwarfed by that from Colombian cocaine, the recent flush of drug money in Jamaica is locally important. In 1984, Jamaica's exports of ganja were estimated at 1,900 tons worth about US $1.4 billion, one-eighth of which returned from the United States to Jamaican growers and handlers. A Jamaican narcotics squad patrols warehouses, sabotages clandestine airstrips, and intimidates growers, and the Jamaican government has sponsored helicopter flights over parts of the island to burn ganja fields. Yet these campaigns from a Jamaican government pressured by US politicians are not appreciated by the majority of Jamaicans; a recent poll revealed that 62 percent of all Jamaicans opposed the curtailment of marijuana exportation to the United

States, in part because so many benefit from it (Treaster 1984). Rank-and-file Jamaicans apparently represent the sentiment of peoples throughout the region because "[e]very Carbibbean politician believes that the drug problem is one of American demand" (Elliott 1988: 17). Many of these same politicans also feel that drug money has the potential to undermine the political stability of the Caribbean, an ironic development that envisions US drug addiction financing the political destabilization of a region vital to United States security.

If the heart of the drug problem is in the United States, its metaphorical veins and arteries permeate the Caribbean. And discussions of the geopolitical dimensions of the problem late in the twentieth century abound with rhetoric concerning the use of drug money to achieve political ends. United States intervention in Nicaragua has been alleged to be associated with drug money as was the anti-US Noriega regime in Panama, and spokesmen occasionally hint that the cascade of cocaine into the United States actually has been orchestrated by Fidel Castro on behalf of international Communism. Frank McNeil, the United States Ambassador to Costa Rica from 1980 to 1983, has stated flatly that the crisis in the circum-Caribbean region is not one of ideology but rather "the capacity of drug cartels to buy governments" (Pitt 1989).

Even academics, whose abstractions often are more disappointing than edifying in comprehending real-world events, can offer perspective in understanding the Caribbean drug traffic into the 1990s. Historians have begun to discuss the comparative roles of the tropical "Big Fix" stimulant commodities – tea, coffee, cocoa, sugar, tobacco, and opium – in the coevolution of developed and underdeveloped parts of the world (Wolf 1982: 332–46). Cocaine is another of these commodities. And devotees of world-economy thinking could hardly find a clearer example than the cocaine trade to demonstrate that a far-flung economic system, once set in motion, is difficult to stop. In producing and transporting narcotics for metropolitan consumption, Caribbean peoples simply are providing tropical staples for external sources, just as they have for the past five centuries.

# 6

## Human migrations

Most of the narcotics passing through Jamaica then smuggled into North America for the United States market apparently are distributed by a clandestine network of US drug pushers already in place. But, late in the 1980s, along with the increase of marijuana and cocaine brought from and through Jamaica to the US, a new breed of Jamaican criminal also has entered the United States. Joseph Vince, an agent of the US Bureau of Alcohol, Tobacco, and Firearms, who is based in Miami, asserts that the Jamaicans are "the most rapidly growing organized (criminal) group in the United States." Within a very few years, these Jamaican men and women – whose numbers are estimated in the thousands – have organized a drug-based crime network on US soil that has extended itself well beyond Miami and New York into the American heartland. During 1987, profits from Jamaican-run "crack houses" operated in Dallas alone were estimated at US $400,000 per day. The Jamaican criminal groups, apparently obsessed with brandishing weapons, refer to themselves as "posses." US drug enforcement authorities attribute the fearless character of members of the newly arrived criminal group as rooted in the impoverishment of their home island, conditions that have given these men and women the attitude that they have, literally, nothing to lose (Cohen 1988).

The stereotypes produced by these Jamaican criminals distress the long-term Jamaican residents of the United States. All Jamaicans know, even if most Americans do not, that tens of thousands of Jamaicans and other black West Indians have emigrated to the United States since early in the twentieth century. Many have contributed substantially to the US becoming leaders in business, politics, and education. That is why most Jamaican-Americans feel slighted by the seemingly slanted news coverage that tends to stereotype all Jamaicans as criminals. Ransford Palmer, the Jamaican economist who teaches at Howard University in Washington, DC, suggests that these inaccurate stereotypes are produced not only by a violence-obsessed news media but by US government officials as well (Pressley 1988).

The Jamaican criminals, it is reported, send substantial amounts of money back to their home island. In this way, and only in this way, they are similar to the thousands of other Caribbean peoples who have emigrated to Europe and North American urban centers in recent decades. Very few of these recently emigrated peoples have lost touch with the places they have left behind, and the probable majority of them send money and material goods back to the Caribbean as a matter of routine. The taxi driver from Barbados in New York, the laundry worker from Aruba in the Hague, the street vendor from Guadeloupe in Paris – not to mention the cane cutter from St. Vincent in Trinidad – all contribute materially to their home societies by remitting gifts or money or by paving the way for friends and relatives back home who may themselves decide to migrate in the future.

In many ways, the current generation of Caribbean migrants is simply doing what earlier generations have done for one and one-half centuries. Since British slave emancipation in 1834, men and women of the region have ventured abroad in order to compensate for the lack of resources and opportunities at home. They have thus broadened their livelihood possibilities by extending their travel patterns ("migrating") through extraordinary and costly individual efforts. Each successive generation of West Indian peoples has identified and traveled to nearby and distant locales. So a successful migration tradition characterizes much of the Caribbean region in the late twentieth century. In some of the smaller islands, human migration sustains the local societies which would collapse without it.

Today's news media take little, if any, notice of the historical background of Caribbean migration, adopting instead an ahistorical slant that inevitably obscures understanding. Massive human movements from West Indian locales are events of contemporary international significance, not simply arcane academic topics, and these migratory movements are duly reported in metropolitan newspapers and on television. The Jamaican drug dealers are only one example. Even the most casual TV viewers are aware of the Mariel boatlift from Cuba to southern Florida, the ongoing "immigration issue" in the United Kingdom, and a perceived political shift to the right in France, said to be partly a reaction to the metropolitan presence of immigrant guest workers from the former French Empire. But a 30 second TV spot on the nightly news describing the close physical proximity of Haitian poverty and Miami's bright lights frames the issues of Caribbean migration entirely in the present; such a perspective ignores the fact that Haitians and other Caribbean peoples have been migrating and returning home for decades. Furthermore, metropolitan perspectives on Caribbean migration often carry with them an obscurantist, condescending, uninformed bias; academic seminars, congressional hearings, or symposia that deal with the "problems" of Caribbean migration thereby hold the assumption that such problems are aberrations. In this case, the norm, of course, is the uniquely bounded, sedentary affluence of Western society.

The root causes of Caribbean migrations are intertwined with the developed world, both now and in the past. Since the 1960s the accelerating magnitude and breadth of international capital investment – mainly from the United States – has helped create a heightened international labor demand and the rise of "global cities" that have attracted, among others, tens of thousands of Caribbean peoples. Investment capital devoted to modernizing Third World areas often disrupts traditional agricultural systems, uprooting people and catapulting them into the wage labor market such as occurred in the Greater Antilles in the early twentieth century. Now these labor movements routinely cross international boundaries, movements facilitated by political decisions such as the relaxation of immigration restrictions (Sassen 1988).

Yet capitalist investment strategies have created local dislocations and incongruities in the Caribbean for many decades, conditions which – combined with the region's fragmented insularity – have helped to produce migration. These incongruities may be said to have begun with the introduced overpopulation that was slavery, and they have created ecological conditions with which Caribbean peoples have had to cope ever since. They also help explain why human migration has been a compelling livelihood strategy in the Caribben region ever since its people have been free to leave.

## The evolution of a regional migration tradition

Escape or the contemplation of escape probably preoccupied the thoughts of Caribbean slaves. This seems to be why birds and boats, obvious symbols of migration, were important among enslaved

Afro-Caribbean peoples (Patterson 1978). Fleeing into the interiors of islands and into the Guianese backlands, a form of resistance discussed in the following chapter, were indeed common during slavery, but actual escape to nearby islands was very rare, although its rarity did not diminish the importance of extra-island movement as an idealized form of escape. Carriacouans still recount a folk legend about a slave rebellion leader on Grenada who swam to safety on Carriacou (Hill 1977: 206). And, throughout the eastern Caribbean, one's ability to see neighboring islands every day must have suggested, at least subconsciously, the possibility of freedom by emigrating prior to emancipation.

The "migration adaptation" that occurred as early as the 1830s after emancipation involved thousands of men and women and was most notable on the smaller islands of the British Caribbean. On Barbados the fertile soil and comparatively level terrain of the entire island remained monopolized by sugar-cane planters; black Barbadian freedmen had few local opportunities other than remaining on estates as poorly paid "located laborers." It is therefore not surprising that many opted to emigrate, often in a semi-clandestine manner, to neighboring Trinidad and also all the way to British Guiana where planters offered higher wages than at home. By 1842, Barbadian colonial officials estimated that perhaps 4,000 Barbadians already had traveled away and that probably 10 percent of those had returned to Barbados. But colonial censuses in Barbados and other Caribbean colonies of the nineteenth century were generally unreliable. We know that migration and return was extensive before 1850 because planters and colonial officials continuously grumbled about it, but the precise numbers of migrants will probably never be known (Richardson 1985: 100–4).

Leeward Islands planters enacted trespassing and vagrancy laws to keep freedmen rooted to the estates. Within months after emancipation, however, hundreds of free men and women from St. Kitts, Nevis, Montserrat, Antigua, and the Windward Caribbean had sailed south in response to wages offered by Trinidadian planters that were double those paid at home. By 1845, more than 10,000 migrants from small West Indian islands had traveled to Trinidad; over 8,000 others had gone to British Guiana. The majority accomplished their journeys on the decks of tiny fishing sloops and schooners over hundreds of miles of tropical ocean interrupted by stormy interisland passages and unmarked rocks and shoals (Hall 1971: 41).

Perhaps the most remarkable dimension of this early migration adaptation was that many of the emigrants eventually returned. Again, accurate records never were maintained. By 1848, however, Trinidadian officials complained that thousands of the "old island-ers" (those from the smaller islands to the north) had gone home. The returnees took money back to friends and kinsmen who had stayed behind. And, as early as 1854, colonial officials on St. Kitts-Nevis reported that travelers had returned from Trinidad arrayed in "gaudy" and distinctive clothing. Successful Caribbean migrants of the mid-nineteenth century thus seem to have adorned themselves back home in modern, fashionable clothing they had purchased at their destinations. This distinctive clothing signified to those staying behind (just as it does among returned West Indian migrants of the late twentieth century) that the migrants had traveled far and prospered abroad. Much more important, these early and hazardous migrations and returns – accomplished in part as a form of resistance against plantation oppression – established a migration tradition in the eastern Caribbean that has endured to the present day (Richardson 1980).

The migration and return of freedmen from the small islands of the eastern Caribbean in the 1840s were intertwined, as they always have been, with world economic trends and adjustments. At emancipation, the center of gravity of the British Caribbean sugar-cane industry was shifting to the south, away from the eroded soils and antiquated infrastructures of the "old islands" toward the "new" southern colonies. The steam engines that had crushed canes in Spanish Cuba before the end of the eighteenth century soon were adopted in the newly acquired British colonies of Trinidad and British Guiana. And the labor shortages perceived by British planters in the southern Caribbean during the new industrial era caused estate owners to send labor recruiters to the smaller islands to seek those willing to emigrate. Eventually southern Caribbean labor needs were met by the importation of the indentured Asian workers enumerated in Chapter 3.

International shifts in capital investment associated with the extension of global trade routes pushed Jamaican workers to Central America in the mid-nineteenth century. A formal agreement between a US company and the government of New Granada (Colombia) in 1848 provided the financial basis for railway construction across the isthmus of Panama. The railway, completed in 1855, was a crucial link in a circuitous route between the eastern US and California prior to the completion of the transcontinental railroad across North America in 1869. Jamaicans, along with Chinese, provided the bulk of the unskilled construction labor force to build the Panama railway. They came to Panama via steamship from Kingston. Possibly as many as 5,000 Jamaicans emigrated to Panama to work on the railway construction between 1850 and 1855 (Newton 1984: 91; Petras 1988: 53–84).

Jamaicans also provided the bulk of the labor force during the failed French effort at constructing a canal across Panama from 1881 to 1888. Few data remain from this movement of workers, but the volume apparently was heavy in both directions between Jamaica and Panama. In late 1884, the Jamaican Registrar General estimated that nearly 35,000 had left Jamaica for Panama during the previous three years and that 15,000 already had returned. The availability of deck passage on steamers from Jamaica, the relatively large population of the island, and its proximity to Central America also made it a logical labor reservoir for the opening up of the banana plantations on the Caribbean's eastern rim. As early as 1888, nearly 2,000 Jamaicans already had emigrated to Costa Rica (Proudfoot 1950: 14–15).

But early Caribbean migration trajectories were not simply directed toward economically bright destinations. Economic dislocations and shifts that created opportunities in some places produced, as often as not, downturns in others. The severe economic depression in the 1880s and 1890s in the sugar colonies of the Commonwealth Caribbean meant lowered wages and grim conditions at home that inspired local residents to seek work elsewhere. So work opportunities in Central America, although they necessitated extended family separations and arduous travel experiences, were doubly attractive to working peoples from all of the British islands during the depression years. Although Jamaicans comprised the bulk of the Antillean labor force for the French canal project in Panama, for example, workers also came from the smaller islands. French-speaking St. Lucians were especially welcome, but when the canal effort was abandoned in 1888, hundreds of black St. Lucians and others were stranded in Central America (Eves 1893: 243).

Nineteenth-century migrations of Caribbean peoples usually were confined to the circum-Caribbean zone, including Central America. But some went farther afield, especially at the end of the century when improved technology extended travel routes. A few black West Indian sailors from Barbados and the Grenadines were harpooners aboard whalers in the Pacific, and William T. Shorey, a black Barbadian, became captain of the Pacific whaler *Alexander* in 1889 (Cohn and Platzer 1978: 88). A tiny number of West Indian migrants had discovered the United States by the turn of the century (Reid 1939). The sugar depression pushed some from Trinidad and nearby islands to seek work in South America, especially as mine workers. One shipload of depression-stricken workers, an estimated 270 men from Barbados, Martinique, and Guadeloupe, even were hauled across the Atlantic after they signed up for railway construction labor in the Congo. None of them returned from the journey (Richardson 1985: 107).

### Twentieth-century migrations

As the political and economic presence of the United States came to overshadow and direct activities throughout the Caribbean early in the twentieth century, the movements of labor migrants there were accordingly associated with – in some cases nearly mandated by – US labor needs. When officials of the Isthmian Canal Commission (the US governmental agency responsible for the construction of the Panama Canal) sought to establish a labor-recruiting terminus in Jamaica in 1903, they were rebuffed by the Jamaican government because too many Jamaicans already had died and suffered working for the French. So the Americans turned to Barbados and, from 1905 to 1913, US officials shipped 20,000 Barbadian male contract workers – as well as hundreds of others from nearby islands – from Bridgetown to Panama (Richardson 1985).

But men and women traveling informally from their home islands to the Canal Zone far outnumbered contract workers during the construction decade. Perhaps as many as 40,000 Barbadians (besides the 20,000 contract workers) traveled informally to the Panamanian isthmus before the canal was completed, and possibly as many as 80,000 Jamaicans did so as well. British West Indians were not the only Caribbean peoples traveling to the Canal Zone. Men and women from Danish, Dutch, and French islands went too. Between 1905 and 1907 US labor recruiters shipped 7,600 contract laborers from Guadeloupe and Martinique to Panama before the continental French government ended the recruiting (Newton 1984: Conniff 1985).

Whatever the total number (insular demographic data were still notoriously vague and unreliable), the West Indians who traveled from their home islands to Panama – and often back again – set in motion demographic trends that reverberated throughout the region. Thousands perished on the isthmus from exhaustion, disease, and landslides; nearly 6,000 Barbadians alone died of all causes from

1906 to 1920 in Panama. Probably over 15,000 British West Indians altogether died in Panama before 1920. Others never returned home but traveled west to the new American banana plantations in Honduras and Costa Rica. Hundreds joined the British West Indies Regiment of World War I and fought against the Turks in Palestine. Thousands stayed in Panama and became the black "Zonians" whose presence would become a treaty issue between the United States and Panama decades later.

The flow of laborers to the Panama Canal Zone in the first two decades of the twentieth century was paralleled by a countercurrent of money that workers sent home to their families and friends. The "Panama money" softened the effects of the economic depression in the British Caribbean. Wives and mothers on the home islands used the wages sent from Panama to purchase foodstuffs, clothing, and membership in local burial societies. In thousands of cases, money from Panama also purchased land plots, fishing sloops, and shops throughout the islands, thereby affording working-class blacks a measure of independence from local plantocracies. This was not the first time small-island West Indians had prospered from money sent and brought home from a wage destination, but Panama money intensified the search for wages abroad by migrants from small Caribbean islands because it represented a volume and continuity of remittances that had never been known before.

Some of the West Indian veterans of work gangs in Panama moved on to the Greater Antilles where American capital was transforming the landscpe and also producing jobs. The attraction of wages in modernizing and constructing new agricultural facilities, and eventually harvesting the sugar cane, in Cuba and the Dominican Republic drew thousands of migrating black "Antillanos". Haitians came too, mainly as seasonal cane cutters. Sailing schooners, and later steamers, had taken men from the English-speaking islands of the Leeward Caribbean to San Pedro de Macorís – in the heart of the Dominican Republic's sugar-cane belt – since before the turn of the century. And Jamaican workers traveled to Cuba, mainly on steamships, in ever larger numbers as the Panama Canal construction wound down. In 1919 and 1920, the peak years for Jamaican migration to Cuba, nearly 50,000 Jamaicans sought work on the larger island. As a linguistically alien, dark-skinned, Protestant, and mostly male labor force, black West Indians were not always welcome in the large Roman Catholic, Spanish-speaking island. In Santo Domingo they were derided as *cocolos* by native Dominicans. Authorities in both countries insisted that these migrants – many of whom arrived for the annual cane harvest in January and departed in July – were holding jobs that should be performed by locals (Knight 1985).

The application and withdrawal of US capital at various locations throughout the circum-Caribbean region thus pulled and pushed labor migrants here and there, affecting the insular demographic patterns that Caribbean migration societies have been known for before and since. Females, children, and old people had tended to predominate in the small islands ever since principally male laborers had traveled away after slavery. But by the early twentieth century insular populations had become even more mobile and fluid from season to season and from year to year as external job opportunities appeared, disappeared, and reappeared. For instance, the massive exodus of men from St. Kitts, Nevis, Anguilla, and Antigua to the Dominican Republic at the beginning of each year in the first decades of the twentieth century began a six-month period in which mothers, wives, and children back home waited, hoping for remittances through the mail. Colonial officials in the same islands dreaded the men's return in the late summer because, for the next few months, unemployment always became a problem on the local sugar-cane plantations, incipient labor protest and disturbances surfaced, and burglary rates rose. The population characteristics, economic opportunities, and even cultural attributes of these small Leeward Islands in the early twentieth century were influenced not so much by local events as by the rhythms of the sugar-cane harvests in the Greater Antilles (Richardson 1983: 129–30).

Inevitably, as more West Indians traveled further from home, they discovered the eastern United States, mainly the New York City area, as a migration destination. Some already had taken up residence in New York and Boston in the late nineteenth century as steamer lines expanded individual travel possibilities. But from 1901, when 520 "Negro Immigrant Aliens" were admitted from the West Indies, until 1924 when 10,630 arrived, 102,000 black West Indians entered the United States. Data were not maintained as to island of origin; some came from French, Dutch, and Danish Caribbean islands, but the great majority were British West Indians (Reid 1939).

The influences of West Indian migrants in the New York area went far beyond the establishment of cultural enclaves in a new environment. Caribbean migrants to New York were particularly vocal and assertive in the early twentieth century, often paving the way for new black professional opportunities that had previously been open only to whites. Influential West Indians, moreover, demanded greater personal respect and established new standards of black self-identity (Anderson 1982). Jamaican Marcus Garvey arrived in Harlem in 1916 and transferred the headquarters of the Universal Negro Improvement Association (UNIA) from Jamaica to New York the following year. The UNIA's newspaper, *Negro World*, indeed had an international circulation; among other things, this weekly stressed black pride and the importance of Africa as the original black homeland. Garvey's influence therefore extended well beyond the Caribbean. For a brief time, he played an extremely important role in the evolution of black consciousness in the United States (Cronon 1964).

The stream of West Indian migrants to the United States was abruptly curtailed on July 1, 1924, when the US national origins immigrant quota law went into effect. In part a reaction to the massive influx of eastern Europeans and Chinese, the new law essentially closed the United States as a viable migration destination for men and women from most of the Caribbean region. During the first half of 1924, over 10,000 black West Indians had come to the United States. In the following year, only 308 arrived. The external sanctions against West Indian migration were thus similar to those that had occurred elsewhere in the past and which would reoccur in the future.

In the Caribbean itself, destinations for migrating men and women were also changing, as macroeconomic controls over local events created boom-and-bust labor markets, sending workers home from some islands and attracting them to others. The high world sugar prices, buoyed by shortages created during World War I, plummeted in the early 1920s. Accordingly, the volume of seasonal labor migration to the Greater Antilles by cane cutters from smaller islands was sharply reduced. At almost the same time, the construction of the petroleum refineries on Curaçao (1918) and Aruba (1929) began to attract thousands of laborers from the British and Dutch Caribbean to build factories, warehouses, roads, piers, and barracks. The construction of the Lake Maracaibo derricks by American engineers had earlier attracted hundreds from Trinidad, Grenada, and St. Vincent – English-speaking labor migrants who had experience working in and around the water – to the large estuarine lake of northern Venezuela.

The economic depression of the 1930s intensified rivalries between host peoples of some Caribbean states and those who had come from neighboring places to work. For years, for example, Dominicans had complained bitterly that migrant workers had taken jobs that locals should perform. In 1929, owing partly to economic depression, the Dominican Republic severely restricted seasonal immigration into the country, a law aimed in part at the sugar-estate laborers who arrived every January from the eastern Caribbean. Tragedy in the Dominican Republic was also related to curbing immigration; it came in 1937 with the slaughter of between 15,000 and 20,000 Haitians who lived and worked in the western and central part of the country. The so-called "Trujillo massacre" was directed by the Dominican dictator and accomplished by those who resented the presence of Haitian migrants who had traditionally crossed over the border to work on Dominican sugar estates (Williams 1970: 466).

World War II provided short-lived relief for labor migrants in the West Indies when the United States took control of the military bases on the several British Caribbean islands. Jobs in Trinidad, Antigua, British Guiana, and also on St. Thomas attracted thousands of West Indians – many sailing from nearby islands aboard schooners – who extended airplane runways, fortified harbors, constructed military barracks, and worked as messengers, cooks, and maids. But, according to older West Indians who recall working for the Americans during the war, these jobs, as so many other jobs had been for Caribbean migrants in the past, seem to have been over nearly as quickly as they appeared. To make matters worse, economic conditions on some of the smaller islands after World War II had deteriorated even from where they stood during the depression. On the British islands, the sudden devaluation of the British pound in 1949 had the immediate effect of increasing prices for items imported from outside the British realm.

A momentous consequence of the aftermath of World War II in Britain – the need for unskilled labor to repair and rebuild the country after a war that had reduced British manpower – led to massive emigration of British West Indians to the United Kingdom in the 1950s. Caribbean blacks traveled to England on British passports so that reliable data were never available as to how many had gone to Britain or had left a particular island; estimates of total Caribbean migration to Britain between 1951 and 1961 vary from 230,000 to 280,000 (Peach 1968). The corresponding loss of people from some of the smaller islands during the decade was astonishing. Montserrat, for example, is said to have lost over 30 percent of its people to Britain during the 1950s (Philpott 1973). The travel itself was accomplished on steamers and charter flights. Husbands and fathers often went alone and established an economic foothold in England before sending for the rest of their families. A disproportionately large number of skilled workers – carpenters, masons, plumbers, electricians – left the islands for higher-paying British jobs, thereby depleting insular work forces and, according to some spokesmen, draining away the most capable and productive local inhabitants.

The Caribbean immigrants to Britain concentrated themselves heavily where jobs were available – in the London area and in the industrial towns and cities of the British Midlands. Far from being greeted warmly in the "mother country," however, British West Indians often were subjected to racial slurs and insults at work and relegated to the worst housing conditions in the unfamiliar British cities. In August, 1958, riots pitting white toughs against immigrant West Indians broke out in the Notting Hill district in West London. Blacks were chased and beaten and some of their flats firebombed in "some of the worst outbreaks of civil unrest and racial violence in Britain this century" (Pilkington 1986: 6).

In the early 1960s, amid widespread white British suspicion and resentment toward an accelerating volume of immigration from the former British Empire, the Commonwealth Immigrants Act was passed. The law, approved after acrimonious parliamentary debate in April 1962, took effect three months later. It specified that those West Indians already residing in the United Kingdom on July 1, 1962, could thereafter bring only wives, husbands, or children under sixteen from abroad to live with them. All others were essentially barred from living in Britain. The British government pointed to a saturated labor market, claiming that unrestricted immigration would create severe unemployment (long a problem in the British Caribbean islands) in the home country. Many West Indians condemned the British action as hypocrisy. And the new British law represented yet another external sanction against Caribbean migrants (Patterson 1969).

Britain in the early 1990s has roughly 650,000 black citizens who continue to be concentrated in the nation's industrial slums. Most black Britons are those who immigrated from the Caribbean three decades ago plus an increasingly restive younger generation born in the United Kingdom. British blacks never have been truly assimilated into white Britain, and a stagnant British economy has sent unemployment rates soaring among its black populace. The general economic disenchantment of Britain's blacks was mirrored in the Brixton riots in south London in April 1981: a large black mob rioted, looted, and confronted local police in disturbances said to be provoked by police harassment. An official report issued in late

1981 outlined the economic and social malaise of Britain's blacks and vowed improvement. Improvement, however, has not been forthcoming, and racial friction continues in Britain's inner cities providing a potent issue for some on the political right.

Unlike the rush to the United Kingdom in the 1950s, the movement of French Antilleans to France was one of modest proportions after World War II. In 1954, those born in Guadeloupe and Martinique residing in France numbered only 15,620, a combined total increasing to 38,740 by 1962. Travel to metropolitan France by citizens of the country's overseas departments is currently facilitated by a quasi-governmental travel agency that also attempts to place new arrivals from the overseas departments in working-class jobs. During the 1970s and 1980s, an increasing number of French West Indians went to the metropole via similar travel arrangements. The 1982 census showed that over 190,000 French citizens residing in metropolitan France were born in the overseas departments of the Caribbean: 9,180 from French Guiana, 87,320 from Guadeloupe, and 94,940 from Martinique. Occupationally, French West Indians in France traditionally have held low-level positions in industry and the civil service. They live mainly in Paris but an equal number is scattered throughout other cities.

The most numerous influx of Caribbean peoples to Western Europe in recent years has been to the Netherlands from the Dutch West Indies, mainly from Suriname. Smaller numbers have come from the Dutch affiliated islands, principally Curaçao and Aruba. Although earlier migrations from the Dutch Caribbean to the metropole were for traditional reasons such as the desire for education, job opportunities, or chances for upward mobility, the high percentage of Surinamese emigrating to Europe in the past decade has been principally because of political push factors. Fearing ethnic rivalry and worse when the South American country gained political independence in November, 1975, tens of thousands of Surinamese of Indian and Javanese descent rushed to Holland in the preceding months (Chapter 8). Early in the 1990s perhaps 200,000 Surinamese and 30,000 Dutch Antilleans reside in the Netherlands.

As in Britain and France, the principal Dutch urban areas have been the destinations of the vast majority of Dutch West Indians, the Surinamese residing mainly in Amsterdam and many Antilleans in The Hague. And similar problems of discrimination and lack of assimilation have been reported by recent West Indian arrivals in the Netherlands, especially as their numbers have increased. In recent years events in both the Netherlands and the Dutch West Indies have also been influenced by the intimate interrelationships between the mother country and its former colonial possessions, interrelationships created by back-and-forth movements of Caribbean migrants. Antilleans in The Hague, for example, demonstrated against the use of Dutch marines to help quell the riots in Willemstad, Curaçao, in 1969, and several West Indian protesters in the Dutch city were arrested and jailed.

Whereas hundreds of thousands of Caribbean migrants and those whose parents are West Indians live in Western Europe, Caribbean peoples now residing in North America (mainly the United States) can be counted in the millions. And in the same way that metropolitan labor needs drew tens of thousands of black West Indians to Britain in the 1950s, a growth in the supply of low-wage jobs in the eastern part of the United States – part of the "restructuring of economic activity in the US" during the 1960s and 1970s – has led to an influx of Caribbean peoples into the United States (Sassen 1988: 56). This influx has been facilitated by the US Immigration Act of 1965 that modified the national origins quotas that previously favored Europeans.

Population estimates of Caribbean peoples in the United States late in the 1980s are confounded by the differences between legal and illegal migrations. Strictly speaking, the United States admits formally 600,000 persons from abroad each year, but some estimates suggest that as many as 500,000 illegal immigrants enter the US annually. The "legal" and "illegal" categories, incidentally, are viewed by Caribbean peoples mainly as bureaucratic hurdles to circumvent rather than immutable law. They are aware of a West Indian presence in the United States since early in the twentieth century, so many in the Caribbean see the changes in the 1965 US immigration law representing continuity rather than change (Carnegie 1983).

Perhaps over five million peoples of Caribbean origin now reside in the United States. Three-quarters of the roughly one million Cubans are in southern Florida, their numbers in the United States augmented by more than 100,000 who came during the Mariel boatlift in 1980. The estimated 800,000 from the Dominican Republic include an undertermined number of "illegals" who have traveled clandestinely to Puerto Rico by boat and then on to the US where they are virtually indistinguishable from Puerto Ricans as far as immigration authorities are concerned (Pastor 1985). Like the Dominicans, the 600,000 British West Indians reside mainly in the New York City area; Jamaicans are a large plurality in this group which includes residential enclaves from every island. The 800,000 (?) Haitians are also mainly in New York, but perhaps 70,000 of them live in southern Florida, many coming by way of the Bahamas (Laguerre 1984: 24–25; Marshall 1979). The roughly two million Puerto Ricans living in the United States are, of course, not "foreign" migrants, although they have often been treated in the same rough way by employers as have undocumented migrants from other Latin American states (Weisskoff 1985: 79).

It would be mistaken to interpret the presence of Caribbean peoples in the US simply as a problem of the cultural assimilation of a newly arrived minority facing "adaptation" pains as it blends into the host society. The size, youth, and locational concentration of the very recent movement of Caribbean peoples north, together with an earlier West Indian presence, have jointly imprinted American culture as well. Perhaps this imprint has been strongest in the New York City area where music, cuisine, fashion, residential patterns, educational expectations, race relations, and much more have been influenced strongly by recent migrants from the Caribbean (Foner 1987: 17–30). Since 1969, each (September) Labor Day in the New York City borough of Brooklyn has seen an ever-growing Caribbean Carnival along Eastern Parkway, whose participants are the recent Caribbean migrants into greater New York. Steel drum bands, blaring reggae recordings, ornamental floats, and exuberant costumes have all been part of the annual celebration. In its growing commercial appeal and ethnic color, Brooklyn's Caribbean Carnival now rivals Trinidad's pre-lenten mardigras festival; unlike in Trinidad, however, the Barbadians, Jamaicans, Trinidadians, and even Haitians, who revel together each Labor Day in Brooklyn have begun to establish, perhaps paradoxically, a Pan-Caribbean identity rarely known in the Caribbean itself (Hill and Abramson 1979).

The United States is not the only contemporary North American destination for Caribbean migrants. The 1981 Canadian census enumerated 172,245 immigrants whose previous habitat was "Caribbean Islands," which, in this case, usually meant the Commonwealth Caribbean. Canadian immigration policies, based on a "point" system stressing education and skills, is said to have drained off many of the more capable and energetic residents from the islands to Canada's benefit. But, according to Canadian anthropologist Frances Henry (1987), black West Indians, regardless of their educational backgrounds, face a rising tide of racism in the cities of eastern Canada where most of them reside. The overall number of black West Indians living in Canada in the mid-1980s was probably twice the number given in census reports because thousands of black Britons with Caribbean backgrounds recently have emigrated to Canada and are officially reported as "British" owing to their most recent country of residence. A Caribbean carnival, with Pan-Caribbean overtones, now also is an annual event in Toronto. Among the roughly 350,000 peoples of Caribbean birth and descent now living in Canada are an estimated 40,000 Haitians, most in the greater Montreal area.

Despite the massive outflow of Caribbean peoples to Western Europe and North America during the latter half of the twentieth century, an even larger number continues weekly, monthly, and annually to travel from place to place within the Caribbean region. The great variety of reasons for these intraregional movements – to obtain part-time work, to seek medical attention, to shop, to sell, to return home, to escape authorities in either one's home or an adopted place of residence – makes it extraordinarily difficult to classify the movements in any meaningful way. Enclaves of recent immigrants inhabit every West Indian state of any size or prosperity; examples include the "down islanders" from the Leeward Caribbean in the US Virgin Islands, Dominicans (from Dominica) in Guadeloupe, and Windward islanders and Guyanese in Trinidad.

These human movements are, in part, spatial expressions of the continuing external control over the Caribbean region, the same control that helped to inspire the earliest intraregional migrations after slave emancipation. Today's Haitian work force in the cane fields of the Dominican Republic represents a tragic case in point. Haitian labor immigration to cut cane in the eastern half of Hispaniola began in the early twentieth century when American occupation of both halves of the island and World War I market demand together heightened export-oriented land use, and thousands of seasonal wage jobs became available in the Dominican Republic (Perusek 1984). Dominicans always resented the intrusion of Haitian labor migrants who nevertheless have continued to provide desperately underpaid labor there, even after the slaughter of the thousands of Haitians by Dominican thugs in 1937. Late in the 1980s, as international financial organizations continue to encourage cash-crop production in both Haiti and the Dominican Republic, and as Haiti's ecological base erodes, thousands of Haitian male laborers are propelled annually into Dominican cane fields. Estimates range from just over 100,000 to half a million annually, numbers that – as in the past – are only vaguely known because of ever-changing labor demands, exploitative non-contract hiring, and changing immigration rules that defy accurate body counts (Latortue 1985). In any case, Haitian cane workers in the Dominican Republic work twelve-hour days, are pitifully underpaid, receive little or no medical attention, and some actually may be locked inside barracks at night. The term *neoslavery* has been applied to these conditions by more than one observer, and the plight of the Haitians working in the Dominican Republic has been the subject of several international human rights investigations during the 1980s. Some American-based human-rights groups have considered petitioning the US government to apply economic sanctions against Dominican sugar imports because they are produced by coerced (Haitian) labor (Plant 1987: 152). But the efficacy of such strategies may be moot owing to the very recent curtailment in the US sugar quota for Dominican sugar which produces desperation not only for Dominican sugar producers but more directly for the drifting army of Haitian canecutters.

### Migration's effects in the Caribbean

An American visitor touring the south coast of the Dominican Republic would doubtless be surprised at the intense and widespread popularity of baseball there. Poor Dominican children use sticks for bats, rocks for balls, and fashion baseball gloves from flour sacks. As they grow older, young men vie for positions on one of the six local baseball teams representing sugar mill towns centered on the sugar refinery city of San Pedro de Macorís (Brubaker 1986). In a similar vein, visitors to the English-speaking Caribbean cannot fail to notice the passion that the local populace displays for cricket. From St. Kitts to Guyana, Sunday afternoons in the countryside often are marked by young men in starched white uniforms engaged in intervillage cricket matches. And when international test cricket is aired live from Pakistan or Australia on local radio and television stations in the eastern Caribbean, local eating and sleeping schedules are adjusted accordingly.

A superficial interpretation might be that the addictions to baseball (the United States' "national pastime") in the Spanish-speaking Greater Antilles, including Cuba and Puerto Rico as well as the Dominican Republic, and cricket, the most British of all sports, in the Commonwealth Caribbean are simply cases of cultural "mimicry," the more important cultures being copied by lesser, peripheral, and essentially colonial peoples. A more satisfactory explanation, however, is that sports in the Caribbean are a metaphor for the outward-looking livelihood strategies (often involving personal migration and return) that experiences with the external world have inspired in Caribbean peoples. They have accepted cultural elements from the outside and adapted them to their own habitats. And in so doing they have changed the outside culture as well. Imagine, for example, county cricket in the UK without its West Indian players or US professional baseball without Dominicans. In a broader sense, international sports featuring the elementary battle between pitcher (or bowler) and batter has provided a competitive arena – without the advantages of race or economic edge – in which Caribbean peoples often have excelled (James 1963: 18).

Yet one need not resort to metaphorical analyses to point out many of the other effects of migration in the Caribbean region. The character of the human populations of Caribbean countrysides, for instance, often is similar to that of Portugal or Sicily; skewed demographic patterns usually show a preponderance of very old and very young people because those of working age have left, permanently or temporarily, to work abroad. And cultural implications of migration-influenced demographies abound. School administrators everywhere in the region, as only one example, lament the presence of rowdy children whose parents are in Canada, the US, or perhaps Trinidad, and whose grandparents cannot control them. Late in the twentieth century both men and women from the Caribbean, not a preponderance of men as in earlier days, emigrate in roughly equal numbers. And it is not unknown in the smaller places for the absence of industrious, working-age peoples to leave behind a certain malaise among the elderly who recall better times in the past when the island bustled with activity (Lowenthal and Comitas 1962).

Caribbean migration demographies are, however, better characterized as volatile rather than skewed, volatility influenced by the external conditions that always have affected Caribbean populations. Indeed, the historically imposed overpopulation in some islands has led to some of the highest human densities in the world. The 1,500 Barbadians per square mile, for example, gives Barbados an islandwide human population density comparable to that of many US suburbs. The fact that nearly all of Barbados's 166 square miles are cultivated in sugar cane gives some idea of how crowded the island's villages, towns, and cities are. And Barbados's relative prosperity is, in part, evidence of how effective migration and return has been as a traditional livelihood strategy there (Richardson 1985: 8). Far across the region, on the eastern edge of Nicaragua's Caribbean coast, the fluidity of the demography of the Miskito

Indians is best explained as a function of traveling away during some periods and returning home in others in order to adapt to altered local economic conditions which are, in turn, affected by external economic cycles (Nietschmann 1979).

As it has so often in the past, money sent home by migrants helps sustain insular Caribbean societies. Whereas a "brain drain" may siphon off many of the best educated of the region, a money flow heading back in the other direction helps cushion the loss. In the past, much of the money sent home was mailed by postal money order and therefore recorded in official government data sources. But in more recent decades Caribbean migrants abroad have sent the majority of their remittances through private banks. Most bankers are reluctant to divulge these data, although it is certain that remittances to the larger Caribbean islands run into the tens of millions of US dollars annually per island (Rubenstein 1983).

Goods and commodities from metropolitan migration destinations also pour into the Caribbean region every day. Material goods from abroad sent and brought back by migrating men and women help to reduce spot shortages of staple items at home but, more often, represent a quality and diversity of commodities otherwise unavailable or prohibitively expensive in the islands. The carry-on luggage on commercial airline flights alone hauled by returnees to Caribbean homelands reveals a wondrous array of goods: blaster radios, toaster ovens, spark plugs, television sets, clothing, groceries, razor blades, and nearly every other item imaginable. The use and display of these goods back home – either by the returnees or their families – signify migration's success, thereby reinforcing its importance in local Caribbean cultures.

The faithful sending home of money and gifts from Caribbean peoples abroad is neither an irrational fetish nor indicative of a preoccupation with the commercial gimcrackery available in metropolitan department stores. Remittances of money and gifts play a functional role for the migrants themselves, especially if external circumstances force them to return. Young men or women who have regularly remitted money in their absence invariably receive warmer homecomings than those who have not. The legendary faithfulness of migrants is, moreover, spread by stories circulated at home, stories that reinforce such faithfulness. Every small Caribbean island has similar tales which often tell of loyal returning migrants sewing money into their clothing to avoid currency regulations or smuggling expensive goods home to benefit their families.

It is not surprising that local prestige often is accorded those who have migrated successfully and returned. In the small British Caribbean islands nearly every shop, taxi, or house of any substance may be traced to the owner or family members having traveled away earlier to a destination where wages were higher and more reliable than at home. Old men who traveled to the Netherlands Antilles from the Grenadines in the 1940s and 1950s returned home to construct elaborate concrete "Aruba houses" with money earned abroad (Hill 1977). "Curaçao houses", named for similar reasons but a different migr̄tion destination, are found on Montserrat (Philpott 1973). Often, rum shops and fishing boats also bear names related to their owners' migration experiences. Many political and business leaders of the Commonwealth Caribbean have resided abroad for extended periods of time.

But support and enthusiasm for those going away is not always unequivocal. The popularity of goods and ideas brought home from abroad, according to some disgruntled observers, can be detrimental to local Caribbean societies, invariably causing young persons to look elsewhere for success and prestige. And a backlash of sorts against those who have emigrated and returned is not unknown. Some of the residents of Suriname have observed with disdain returnees parading through tropical Paramaribo sporting winter European clothing so that no one can mistake that they recently have returned from Amsterdam. In the towns and villages of Puerto Rico, the term "Nuyorican" is not always used to connote awe and respect. Older Barbadians recall with amusement the images conveyed by "Panama men" who returned from the Canal Zone with a strutting self-consciousness, bedecked with the latest fashions from Colón. The character of Caribbean migrants has even become grist for contemporary international propaganda. The Cubans who arrived in Florida in 1980 came via "a freedom flotilla," according to American spokesmen, because they were fleeing Cuban Communism. Cuban officials, on the other hand, designated the movement as the "scum shuttle" to alert the world that those on board the Mariel boats were selfish, lazy, and the dregs of Cuban society.

Migration even influences personality traits in some Caribbean areas. In the islands where men and women have had to migrate and then return in the wake of changing global conditions, personal characteristics of picaresque resilience and economic underspecialization have paid dividends. The outside world has presented an ever-changing array of obstacles, hazards, and rewards. And successful exploitation of the outside world has called for physical strength in some cases or the ability to haggle successfully with immigration officials in others. Economic specialization has held few long-term rewards. In the face of continuing uncertainty, migrating men and women have usually foregone permanent commitments in any one direction except for eventually returning home (Richardson 1983: 171–82). Because coping successfully with the variety of hazards facing these migrants requires them to display many talents to survive, those who return are understandably proud of their achievements, a pride that some interpret as bravado. Along the island arc of the eastern Caribbean, men travel aboard rickety sailing schooners from island to island smuggling whisky, an occupation requiring a good deal of courage in the face of danger. For these men, "A man is either brave or cowardly, and a coward only invites disdain. The desire to be thought of as brave is so strong that expressions such as 'please' and 'thank you' are seldom used because they indicate weakness" (Beck 1976: 41). Similarly, in the English-speaking island of Providencia (owned by Colombia) men who have traveled away prefer one another's company to that of others and consider themselves courageous, successful, and brave: "So when men gather together in rum shops or beneath the palm trees on the beach much of their conversation is taken up with stories of their exploits – mostly those that occurred abroad" (Wilson 1973: 155).

### The drift to Caribbean cities

Not all residential movements since emancipation have involved extra-island migrations. As in nearly every other part of the world, there has been a steady rural-to-urban drift in the Caribbean involving millions of peoples. Their movements, caused by a complex variety of reasons, have left the countryside throughout most of the region less heavily populated than before. In the last decades of the twentieth century, these movements have accelerated in some Caribbean places, leading to the same urban problems of overcrowdedness, pollution, anonymity, and crime familiar to residents of developed areas of the world. Movements from countryside to city in the Caribbean, further, often represent intermediate changes in residence because in many cases they precede migration abroad for individuals and families.

The largest and most heavily populated islands, not surprisingly, have the largest cities, with Havana (2 million) and Santo Domingo (1.4 million), the largest, and with San Juan, Kingston, and Port-au-Prince not far behind (Table 1). Also, the larger islands are the only ones capable of sustaining medium-sized secondary urban areas such as Ponce in Puerto Rico or Santiago de Cuba, Camaguey, and Santa Clara in Cuba. In the Commonwealth Caribbean only Montego Bay in Jamaica and San Fernando in Trinidad could be classified as "second cities." Elsewhere in the English-speaking islands, as in the rest of the region in general, there usually is only one true urban area per island, a city or large town that contains the seat of government, all the important commercial and communication facilities, and the island's premier harbor. These primate cities usually have given rise to conurbations, smaller towns and villages formerly outside the city becoming absorbed into the larger urban area. Rapid growth and an absence of planning around some of these sizable conurbations, such as Pointe-à-Pitre in Guadeloupe or Bridgetown in Barbados, can make morning and evening automobile traffic densities wondrous to behold.

Caribbean cities exhibit great variety. The largest Caribbean urban areas are segmented into identifiable industrial, commercial, and residential districts or suburbs. In contrast, in the places with smaller populations, the capital "cities" continue to satisfy the stereotyped settings reminiscent of mid-twentieth century novels: Cayenne, which has a population of 35,000 – or nearly half the people residing in French Guiana – is a cluster of "generally wooden structures with tiled or rust colored corrugated iron roofs . . . It is only now that a few scattered concrete blocks are beginning to alter this idyllic small-town atmosphere" (Drekonja-Kornat 1984: 27). In the tiny urban places of the Caribbean, such as Hillsborough on Carriacou or Kralendijk on Bonaire, you are as likely to meet a goat or chicken on the street as another person. Furthermore, traditional differences between "urban" and "rural" are confounded on very small tourist islands such as St. Maarten where the village of Philipsburg, the island's main settlement, could hardly be called a "city" compared with the modern, tourist-oriented resorts elsewhere on the island.

Although they may appear ramshackle and old-fashioned to Europeans or North Americans, Caribbean cities can be magnets for rural peoples. To the inhabitant of Guyana's Essequibo coast, for instance, the eight hours by bus (involving two ferry-boat crossings) to Georgetown is time well-invested in order to savor the vehicular traffic, pedestrian congestion, and concentration of cinemas in the capital city where everything is "bright, bright." Indeed, in the Caribbean as in other Third World areas the "bright lights" pull theory accounts for many of the rural-to-urban migrations both now and in the past, attractions that have been intensified by Western radio and television advertisements of the good (urban) life: "The city is the shop window of a wanted world; the crowd gathers round it to see and be seen" (Cross 1979: 121). Yet push factors are equally important in explaining rural-to-urban demographic shifts in the Caribbean. Technical changes in sugar-cane processing and milling, replacing earlier labor-intensive activities with factory-like operations, have created rural unemployment and sent erstwhile estate workers to seek jobs in local cities. In the 1950s and 1960s the encroachment of foreign bauxite-mining companies on lands traditionally held by rural Jamaican subsistence producers uprooted the latter and sent them to Kingston by the thousands (Beckford 1987: 16–17).

Beyond the veneer of shop windows and the physical amenities of piped water, indoor plumbing, and reliable electricity, Caribbean cities traditionally have represented physical refuges from the ardor, discipline, and class structure characteristic of the rural plantations. Ever since slavery, when light-skinned freedmen located non-plantation jobs in the urban areas, the cities, not the countrysides, usually have seen what little upward social mobility the Caribbean region has offered its working-class inhabitants. The few white-collar jobs are located in the cities. Accordingly, social contacts through fraternal lodges, beneficent societies, reading clubs, and prayer groups, while not absent in rural areas, are more readily available in the city.

The opportunity for better education is a major attraction of Caribbean cities (Cross 1979: 122–28). Trinidadian economist Lloyd Best has identified education as the "escape hatch" for working-class West Indians of the Commonwealth Caribbean with the local school headmasters traditionally exerting a great deal of control in selecting those few named to go abroad for university education. The presence of the University of the West Indies campuses in Jamaica, Trinidad, and Barbados and its extra-mural centers in all of the smaller islands, has diminished somewhat the aspirations for education abroad. Yet education possibilities still are infinitely greater in Caribbean cities than in the rural areas where the elementary schools are desperately overcrowded and underfunded. Only the brightest boys and girls in the rural districts of the English-speaking Caribbean pursue further education in the best secondary schools which are all located in urban areas. And the tenacity among West Indian school children for education has carried over into the United States where school children of Caribbean parentage often outperform others. Black West Indian school children in England, on the other hand, have had relatively little academic success. Their relative failure is often blamed on teachers' low academic expectations of them, although there is little agreement as to why the children have performed poorly in British schools.

The city concentrations of better educational facilities is important in the Spanish-speaking Caribbean as well. Even in the low-income shantytowns in the San Juan area, "[e]ducation is stressed as the principal avenue to upward mobility." Boys in the shantytowns aspire to jobs as artisans, factory workers, or service workers, jobs available through education. And girls in the poorest areas of San Juan, although traditionally encouraged little as far as education is concerned, often have performed better than have their male counterparts in school, owing to the greater array of job opportunities for females in San Juan than in the Puerto Rican countryside (Safa 1974: 56–57).

Caribbean social cleavages are magnified in urban settings as the prosperous few and the many poor, plus a moderate number in the middle, interact daily. Residential segregation is most apparent in the largest cities such as San Juan where the urban shantytown districts are far removed from the prosperous districts of Santurce and Río Piedras, formerly self-contained communities that have become part of the greater metropolitan area. In smaller Caribbean cities, such as San Fernando, Trinidad, where there is noticeable rivalry between "Creoles" (those of African background) and the descendants of indentured laborers from India, a smaller city size – providing a heightened possibility of ethnic interaction – intensifies these rivalries (Clarke 1986).

The concentrations of working-class peoples in Caribbean cities have inevitable sociopolitical implications. Where the same people share residential habitats, workplaces, religious identities, and memberships in various voluntary organizations, as they do in Caribbean cities, a capacity to organize is a crucial consequence of these "many links that extend into the larger society" (Wolf 1982: 360). And protests are much more successful if they are joined by the number of people a city can provide. A general strike in St. George's, Grenada, early in 1951 was instrumental in bringing Eric Gairy to the forefront as a new political leader on the island (Singham 1968: 152–69). Similar examples abound from the other islands. There also is evidence that political protest in the Caribbean, drawing its energy from an urban setting, can melt away without that setting. Early in 1970, the "Black Power" disturbances in and around Port-of-Spain, Trinidad, were energized by a hard core of unemployed young urban blacks. The leaders of the movement, consciously attempting to extend the disturbances into the countryside, organized a march into the sugar-cane belt of western Trinidad, only to be greeted with indifference and even hostility by the East Indians there (Bennett 1989: 136–39).

The occasional sociopolitical outbursts in Caribbean cities reflect unemployment levels reaching 30–40 percent because Caribbean urban areas do not absorb redundant rural wage labor any more than local Caribbean businesses are independent of global economic decision-making. As in other Third World areas, working-class residents of Caribbean cities often are underemployed, underspecialized, and ideal candidates for the ephemeral wage jobs provided by itinerant assembly plants catering to the US market. It is little wonder that thievery, assault, and prostitution thrive in many of these places. Even in Cuba, whose urban growth has been controlled by a government whose priorities have been in developing the countryside and whose urban housing districts are of generally high standards, the crime that does exist usually is associated with the urban areas. Groups of "hippy" youths are said to congregate in the La Rampa area of Havana and to indulge in various acts of vandalism, theft, and drug use (Salas 1987: 342).

For most, movements from the boredom of the Third World countryside necessitates an immersion into the precariousness of an urban existence. The classic Caribbean case is the young man from an interior hillside village of Jamaica who seeks fame and fortune by moving to Kingston. In the city he soon finds himself residing in one of the squatter shacks in West Kingston built on government land, a one-room hut constructed of cardboard, packing crates or discarded burlap bags. His economic pursuits consist of a variety of legal and illegal activities, including the salvaging of small bits of food, housing material, or saleable items from the "dungle" or city dump (Clarke 1983). The dead-end character of this existence is reinforced daily when he sees prosperous Jamaicans and occasional tourists who represent the good life. His despair would have plunged to abysmal depths on September 12, 1988, when hurricane Gilbert transformed Kingston shanty towns into "a tangle of fallen trees, phone lines, and twisted aluminum roofs . . . [creating] a desperate shortage of fresh food and water" (Engardio *et al.* 1988: 32). Then one day he is approached by a well-dressed narcotics smuggler ablaze with a gold necklace and fragrant with after-shave lotion offering a forged passport and stories of unparalleled glamor, riches, and excitement in the United States. A choice between the hopelessness of living on the dungle of West Kingston and membership in a US drug posse? It would require only the most superior form of ethnocentrism not to understand which way such a young man would choose.

# 7

## Resistance and political independence

At the beginning of World War II, in late 1941, American concern over the safety of Suriname's aluminum ore led to the occupation of the Dutch colony by 1,000 US troops (Baptiste 1988: 115–29). The main fear was of a possible commando raid out of Vichy-held (pro-Nazi) French Guiana to sabotage the bauxite mines at Moengo in eastern Suriname, or even the blocking of the narrow Cottica River leading to the mines at Moengo. Accordingly, a small concentration camp near Paramaribo in Suriname housed suspected German agents and also German ship crews for the duration of the war. In 1942, three Germans escaped from the camp and traveled overland in an easterly direction to the Maroni River separating Suriname from French Guiana. Upon attempting to cross the river, the Germans were recaptured by the men from a "Bush Negro" village and returned to the Dutch authorities in Paramaribo (Sharp 1942: 14).

The black villagers who captured the German escapees probably were members of the Djuka tribe, one of six Maroon ("Bush Negro" from the Dutch *bosch neger*) tribes inhabiting the forested interior of Suriname. These peoples, who until very recently represented over 10 percent of Suriname's human population, all descended from escaped African slaves. In the late seventeenth and early eighteenth centuries slaves fled inland from the coffee, timber, and sugar cane plantations on Suriname's coastal plain to establish clandestine encampments in the rainforest. Then for over a century they and their descendants fought a war of liberation against European soldiers. In 1762, one hundred years before the general emancipation of slaves in the Dutch Caribbean, the Maroons of Suriname won their own freedom and autonomy from Dutch control. In the decades thereafter, they have inhabited scores of villages located along the

158

banks of Suriname's interior rivers. In dress, language, and material culture, the Maroons of Suriname exhibit strong resemblances to these same elements of the West African civilization from which they were torn and sold into slavery. They are mindful of their African heritage, although they reckon their establishment as a people in their escape from slavery. This identity has been maintained through a rich tradition of dance, music, and oral history commemorating their escape from and subsequent victory over the representatives of a European plantocracy. The veracity of this folk history, moreover, has been authenticated through the brilliant research of anthropologist Richard Price who has corroborated the oral history of the Maroons of Suriname with archival records housed in the Netherlands (1983). Sadly, after more than two centuries of an autonomous existence in the Suriname rainforest, the descendants of the Maroons may now be eventually disappearing in Suriname's little-known civil war in which the "Bush Negroes are dying in a situation close to genocide" (Brana-Shute 1987: 28).

The Maroons of Suriname are a living testimony to the resistance among slaves that marked every slave society of Atlantic America, from Brazil to Virginia. Runaway slave societies existed throughout the Caribbean region, their success and size heavily dependent upon the presence of geographical refuges adjacent to the plantations themselves. In the Guianas, with upriver rainforests, and in the Greater Antilles, with large and sometimes impenetrable interiors, Maroon societies of remarkable size and tenacity emerged. Maroons erected elaborate fortifications, provided their own subsistence, developed meaningful cultural practices and, in some cases, traded actively with the plantations societies from which they had sprung. Small-scale insularity did not reduce the slaves' antipathy to slavery because there were runaways and incipient Maroon societies on every island.

The significance of the Maroon societies of the Caribbean extends beyond individual islands and beyond the era of slavery. In the late eighteenth century, in Haiti and to a lesser extent in the Windwards, Maroon wars and parallel clashes between European forces in the Caribbean became intertwined. European colonization strategy was therefore affected and, in some cases, thwarted by Maroon uprisings. Even more important, the Maroon tradition is not forgotten in the Caribbean. In the interior of Jamaica, as well as in the back country of Suriname, living Maroon communities continue to reckon their descent from those who escaped from coastal plantations. Increasingly, school history lessons in the Caribbean celebrate the exploits of Maroon leaders, as well as those of Columbus, Napoleon, and Washington. In several places in the region, statues and monuments commemorate Maroon accomplishments, such as in Guyana where a statue of the Maroon leader Cuffy – who led a rebellion against the Dutch in 1763 and held much of what is now eastern Guyana for nearly a year – stands outside the residence of the Guyanese president.

The resistance to oppression by Caribbean peoples has continued until the present day. The remarkably harsh character of Caribbean history – marked by ecological destruction, human expendability, and coercive labor systems all orchestrated by external powerholders – has been met throughout history with resistance by the region's common people. The most dramatic example is Haiti where rebellious slaves defeated metropolitan armies and established the second oldest independent state in the Western Hemisphere. Guerrilla warfare in the region, however, has continued well into the twentieth century, the most vivid example that of the Cuban Revolution in the 1950s. As organized resistance to outside control of the Caribbean became more "legitimate," moreover, its locus often changed from mountain strongholds to the streets of Caribbean cities and villages where the oppressed rioted and demonstrated for greater control over their own destinies, as occurred in the British islands early in the twentieth century.

This chapter emphasizes the overt dimensions of the resistance of Caribbean peoples, including armed struggle and confrontation. But resistance takes many subtle and less dramatic forms that are often masked by cultural display (e.g., Scott 1985). Caribbean resistance has taken many forms, including emigration, mockery of planters, "laziness," and the establishment of independent village settlements. So, although resistance always has been an important dimension of Caribbean cultures, it does not follow that Caribbean peoples are offensive and truculent because of it. Quite the opposite: the warmth and generosity of the region's peoples are legendary. Despite their warmth and outwardly easy-going nature, however, peoples of the region share a tradition of resistance with others in the Third World. In the words of Immanuel Wallerstein: "The mark of the modern world is the imagination of its profiteers and the counter-assertiveness of the oppressed" (1974: 357). His remark is only partly apt in describing the character of the common peoples of the Caribbean region, for they have resisted domination and dealt with other problems for centuries with considerable imagination and creativity.

## Resisting slavery

The earliest resistance to enslavement in the Caribbean was not by imported Africans but by the region's aboriginal peoples. Though unable to withstand the onslaught of imported disease, the complex of sophisticated armaments, and the diplomatic treachery of the European invaders, the Arawaks, Caribs, and other aboriginal peoples of adjacent coastal zones often fought until they were eliminated, absorbed, or relegated to out-of-the-way "reserves." Aboriginal resistance occasionally was of a regional or interisland character: early in 1511, a rebellion among the Arawaks of western Puerto Rico was coordinated with a diversionary attack on eastern Puerto Rico by the Caribs from St. Croix, an incident leading to scores of Spanish deaths and followed by severe Spanish reprisals (Watts 1987: 106).

Nearly as soon as the first enslaved Africans arrived in the Caribbean – during the first decade of the sixteenth century on Hispaniola – several of them escaped into the hills, beginning a Maroon tradition in the Caribbean that has endured for nearly five centuries (Price 1979: 1; Sauer 1966: 206–7). It is not surprising, furthermore, that early runaway African slaves sometimes allied themselves with aboriginal peoples in common resistance to European domination. But the range of relationships between these two oppressed groups varied greatly throughout the region. In St. Vincent and along the Miskito coast of eastern Nicaragua, for example, a mixture of people of both aboriginal and African descent emerged. In other cases, European planters employed aboriginal woodsmen to track down runaway slaves (Craton 1982: 62–63).

As Caribbean slave societies "matured," Africans' resistance to slavery took on many configurations. The more overt types of resistance, perhaps exemplified by the image of slaves brandishing cutlasses and racing through blazing cane fields, were much less common than daily resistance. Balking, malingering, feigned stupidity, self-inflicted wounds, even suicide were ways in which slaves withheld labor from planters. And "good" behavior among slaves was occasionally a short-term response to slavery with very different long-term goals in mind; in more than one case, trusted house servants or slave foremen took active leadership roles in slave uprisings. The means by which plantation slaves settled scores short of open rebellion were numerous, perhaps poisoning of planters and their animal stock being among the more sensational types. On Martinique in the eighteenth century, the poisoning of livestock by slaves was so common that local whites habitually attributed all animal deaths to slave poisoning and therefore attempted no remedies whatever even if animals fell ill (Tomich 1976: 108).

The term *Maroon* (*marron* in French) apparently comes from the Spanish *cimarrón* that originally was used to designate runaway cattle on Hispaniola (Price 1979: 1). And running away by slaves (*marronage*) was not always permanent; slaves occasionally indulged in *petit marronage*, especially in the French colonies, absenting themselves from the plantations for a variety of reasons and receiving relatively minor punishments on their return. In the British Windwards, "marooning" eventually became a synonym for getting away from it all; in August, 1893, six decades after emancipation, a local colonial official on St. Vincent described the recently improved pathways to the edge of the Soufrière volcanic crater, trails that enabled groups of Vincentians to enjoy "their 'marooning' (i.e., picnic) parties in order to enjoy the wonderful and unique sights provided by nature" (Colonial office [London], 321/149 "Tour Through Island" 24 August 1893).

But "marooning" during slavery was hardly a playful exercise. The savage punishments usually meted out to those permanent runaways who were recaptured by white planters were enough to dissuade any thoughts of surrender. Captain John Stedman was part of a contingent of European soldiers recruited to fight against bands of Maroons in eastern Suriname in the late eighteenth century. When Stedmen observed with disgust the hanging of captured Maroons, and one broken alive on the rack, he was apprised of even worse punishments by a seasoned onlooker, including dismemberment and being hanged alive by an iron hook through the ribs (Stedman 1988: 102–5).

Types of *marronage* seem to have differed significantly among different types of slaves. Escape into the mountains or forests was accomplished by disproportionately large numbers of recently introduced, unacculturated Africans, men and women repelled by the horror of capture in West Africa and the suffocating, manacled voyage across the Atlantic. Their flights into Caribbean backlands often were marked by obviously futile attempts to find their way back to their African homelands; African means of subsistence and survival therefore were important, if modified, in their coming to grips with their new surroundings. Slave runaways who had been born in the Caribbean, and who were thereby escaping from the long-term cruelties of the plantation system, were valuable members of Maroon encampments because of their familiarity with white pursuers. These "creole" slaves also, because of their linguistic abilities and seasoned outlook on the overall society, were better able to lose themselves in urban populations among black freedmen, especially during the latter years of slavery (Price 1979: 24–25).

Maroon settlements were based upon concealment and inaccessibility. Naturally, forbidding terrain was ideal for Maroon villages, and heavy forests, swamps, or steep slopes intensified the already formidable ecological problems faced by runaway slaves in adapting to these environments. In the Guianas, Maroons devised false paths – often leading to fatal quagmires or pit traps – to enhance inaccessibility. The most famous environmental backdrop for Caribbean Maroon settlements was the differentially eroded and hilly limestone landscape of central Jamaica which – although geologically unlike the volcanic interiors of the Windwards – represented a similarly ideal physical sanctuary for Maroons and provided opportunities for the ambush of pursuers. In the interior of Jamaica in the early eighteenth century, the Maroon leader Cudjoe established his settlement that could be approached only by a narrow, uphill trail then a gorge one-half mile long; Cudjoe's men rolled boulders down either side of the gorge to repel invaders, if any got that far (Craton 1982: 84).

Despite the necessity of establishing themselves in wild terrain, Maroon societies of the Caribbean became remarkably skilled in provisioning themselves. Hunting and fishing, naturally enough, were vital activities as well as learning to fashion dwellings, clothing, and tools from materials provided by the immediate environments. And the Maroons often maintained thriving agricultural grounds, a daunting task because of the necessity of locating their village areas in rough terrain and also the need to avoid surveillance by outsiders. At the end of the eighteenth century in the back country of Guyana one of several Maroon encampments was surrounded by a water-filled moat lined with sharpened sticks; but after gaining access to the village area via a partially submerged path, intruders found an assemblage of huts surrounded by cultivated "oranges, bananas, plantains, yams, eddoes, and other kinds of provisions" (Smith 1956: 10). Similar agricultural grounds were common to nearly all the Maroon settlements yet their inhabitants could not produce metal implements such as pots, axes, firearms, and ammunition. So, in order to acquire materials necessary to survive outside European-dominated plantation societies, Maroon settlements were, in part, dependent on these same plantations societies through theft, smuggling, or even trade (Price 1979: 12–13).

European military campaigns to eradicate Maroon villages concluded, more often than not, unsuccessfully, the soldiers having been eluded or encountering abandoned campgrounds. When fighting did occur, the Maroons – who almost always were outnumbered – employed the classic guerrilla military tactics successful among people of Vietnam and Afghanistan in repelling outside armies of superior size and strength late in the twentieth century. Maroon fighting units, intimately familiar with their surroundings, routinely ambushed and decoyed larger aggregations of troops, capitalizing on their ability to maneuver and always on the element of surprise. The Maroons also had much more at stake than did the outsiders, and there is evidence that Maroon cultural practices and military tactics were bound up with one another. Throughout the Caribbean region, Maroon warriors were subjected to complex religious rites in order to render them bulletproof and thereby indestructible from European firearms (Price 1979: 10).

On March 1, 1739, the Governor of Jamaica signed a remarkable treaty with the Maroon general Cudjoe. The latter's followers were declared forever free, and they were given autonomy over a 1,500 acre portion of central Jamaica with the proviso that they could cultivate any crop other than sugar cane. The Maroons agreed to return slaves and to help if Jamaica came under foreign invasion. The fifteen-article treaty did not apply to all the Maroons in Jamaica, and some of the slaves at the time evidently were dismayed by the return-of-runaways clause; but it marked the first act of collective emancipation for formerly enslaved blacks in the Caribbean, freedom that they had won through successful armed resistance against the Jamaican plantocracy (Craton 1982: 89–92).

The treaty was all the more remarkable because it conceded at least a small segment of Caribbean resources, specifically land, to former slaves. As the geographer David Lowenthal notes: "The aim of runaway slaves was more than flight, it was recognized sovereignty over the land itself" (1961: 4). Indeed, it is important that some of the early slave revolts, as distinct from Maroon engagements, had as their ultimate objectives not only individual and collective freedom but the domination of territory and, in some cases, entire island economies. The latter possibly was the objective of the intricately planned but aborted slave rebellion on Antigua set for October, 1736, when Court, Tomboy, and other slave leaders planned to blow up the building in the capital town of St. John's in which the leading planters of the island were celebrating the coronation of George II (Gaspar 1985). Twenty-seven years later, during the great slave rebellion in Berbice (now part of Guyana), the slave leader Cuffy referred to himself as "Governor" and offered to split Berbice in half and share it with the *de facto* Dutch colonial governor (Lewis 1983: 226). Perhaps the ultimate territorially based self-confidence and – to Europeans – brazenness, came from the Maroon leader Quashee in Dominica who, when told the governor had put a price on his head, offered his own monetary reward for the head of the governor (Marshall 1976: 29)! And, when the rivalry between the Black Caribs (An African-Carib admixture) and British over St. Vincent's lands became too intense, the British Navy's solution was to export roughly 1,700 Black Caribs to the Caribbean coast of Central America, an event that occurred in 1797 (Gonzalez 1988: 39–50).

After the British abolished their Atlantic slave trade in 1807, the overall objective of Caribbean slave rebellions seems to have shifted from local take-over to a realization of the personal freedom they felt already had been granted. A major rebellion among the slaves of British Guiana in 1823 was directly attributed to concurrent debate in the British parliament over the future fate of slavery and the associated belief among Guianese slaves that freedom already had been granted by the Crown but withheld by local planters (Smith 1962: 35–37). Similarly, the great "Christmas Rising" in Jamaica in 1831 involved 20,000 slaves, destroyed a score of plantations on the western end of the island and, most significantly, was paralleled by the parliamentary debates in London concerning the slavery issue (Genovese 1979: 36–37). Freedom's imminence in the early nineteenth century in the Caribbean inspired enslaved blacks to resist slavery with an even greater intensity than they had before, and their resistance, in turn, helped in no small way to pressure European powerholders to bring the institution to an end.

### The Haitian revolution

The attitudes of early-nineteenth-century British Caribbean slave-holders and slaves, however, were perhaps influenced even more by the ultimate slave rebellion that just had taken place. From 1791 to 1804, the slaves of the western half of Hispaniola galvanized themselves into an overpowering fighting force, defeating British, Spanish, and French armies, and thereby forcefully transforming French St. Domingue – one of the richest colonial prizes in the world – into the black republic of Haiti. The carnage had occurred almost within earshot of Jamaica. And British policymakers, who early on during the Haitian revolution recognized an opportunity to annex St. Domingue, soon found themselves hoping that the Haitian example would not spread and thus destroy their own Caribbean slave colonies.

"Few wars have been so completely destructive as was the Haitian Revolution" (Ott 1973: 190). And few have ever been so influential for Caribbean peoples. The human population of Haiti fell from an estimated 500,000 prior to the revolution to half that at the end. The colony's plantations, which in the 1780s had provided perhaps half the sugar and coffee consumed in all of Europe, were destroyed. The French army, fresh from victories over the finest soldiers of Europe, were decimated by yellow fever and Haitian forces. And white planter families from the Guianas to Cuba to South Carolina now scrutinized their black fieldhands with a heightened sense of dread. The Haitian revolution, further, possibly reoriented the geopolitical history of the Western Hemisphere by denying the French a springboard for expanding their influence further in the Caribbean and North America (Geggus 1989: 43).

St. Domingue's eighteenth-century renown had not been its potential for slave rebellion but its fabulous agricultural prosperity. Almost unsettled until the end of the seventeenth century, the colony had lagged behind its neighboring territories in plantation development and had not undergone the intense overcropping and soil exhaustion that, ironically, is Haiti's ecological hallmark today. As late as 1788, a visiting delegation from Jamaica noted that St. Domingue's soil fertility explained its high sugar-cane yields and, therefore, the French planters' ability to undersell the British in the European sugar market (Williams 1970: 239). On the eve of the revolution, St. Domingue had hundreds of sugar cane, cotton, and indigo plantations that dominated the colony's lower elevations and nearly two thousand coffee plantations that in recent decades had been hewn from the intermediate, forested elevations (Ott 1973: 6).

The rapid plantation development of St. Domingue in the eighteenth century called for the influx of thousands of people, mainly West African slaves. In the last years prior to revolution, an incredible 30,000 slaves were coming annually to the colony via legal means. In 1789, St. Domingue had a human population that numbered over one-half million: 40,000 whites of varying status at the apex of the colony's social pyramid; 28,000 mulattoes (mixed-blood) and free blacks in the middle; and at the base an estimated 452,000 black slaves, many recently brought to the Caribbean from West Africa (Williams 1970: 246).

The recent arrival of so many blacks into St. Domingue combined with their remarkably diverse cultural backgrounds, including a wide linguistic diversity, perhaps accounts for the surprisingly slight amount of Maroon activity there prior to the revolution compared with the British and Dutch islands. The encroachment of coffee estates into the colony's highlands also was systematically eliminating forested sanctuaries, thereby reducing *marronage* as a viable alternative to plantation slavery (Geggus 1989: 24–25). Yet slave resistance on St. Domingue was not unknown. In 1700, several hundred revolted at Le Cap François, on the north coast, and the leaders fled to the Spanish half of Hispaniola (Ott 1973: 18). Half a century later the Maroon leader Macandal was burned alive as punishment for plotting the poisoning of hundreds of white planters and their black sympathizers in an aborted revolt intended to drive all of the whites from St. Domingue (James 1980: 20–21).

The colony's social complexity meant that disgruntled poor whites (the so-called *petits blancs*) of St. Domingue and especially the mulattoes, rather than the enslaved black masses, were the ones first influenced by the revolutionary winds of change emanating from France in the 1780s. The *grands blancs*, or large-scale planters of St. Domingue, considered themselves a cut above the white colonial administrators who, in turn, looked down on the *petits blancs* who were overseers, artisans, and minor professionals. The mulattoes were becoming more numerous and more prosperous – mainly from St. Domingue's coffee boom – by the 1780s, and already they had come to dominate the colony's rural police force as well as its standing militia. Perhaps the greatest influence on St. Domingue's mulatto group had come from the experiences of some of them in a special regiment that had fought alongside the rebels in Georgia during the US War of Independence. They returned to St. Domingue "with military experience and a new sense of their own importance," and they also brought back a heightened disdain for the French colony's ongoing color caste system (Geggus 1989: 25).

By 1790 mulatto-white rivalries and hatred in St. Domingue were fanned by radical pronouncements from Paris as well as by sociopolitical developments in neighboring French Caribbean colonies. Several mulattoes had been murdered by whites in Martinique in the summer of 1790, and the ensuing riots there called for French troops from Guadeloupe. Brown-skinned Vincent Ogé returned to St. Domingue from Paris that summer, intending to realize political equality for his group with the colony's whites; he inspired local mulatto uprisings in the southern part of the colony but was hanged the following March. In that month French soldiers arrived in St. Domingue to reinforce the largely mulatto bureaucracy, but they soon mutinied and joined the *petit blanc* faction. To complicate matters, news from Paris describing varying political stances toward human rights were interpreted in St. Domingue in many ways. The "May Decree" from the French National Assembly in 1791, for example, called for the enfranchisement of certain St. Domingue mulattoes who met strict property qualifications, but the decree enraged local whites who thought it included all mulattoes (Ott 1973: 35–42).

On the night of August 22, 1791, the slumbering black slave populace – awakened by two years of mulatto–white wranglings and skirmishes – exploded into violence on the colony's northern plain. That night a heavy late-summer storm had drenched St. Domingue so the slaves carried torches to light their way. Their leader was Boukman, a fugitive slave from Jamaica who also used religious incantations to tap deep cultural roots of black discontent. The subsequent murder, burning, and pillage took most of the white planters by surprise, an error in judgement that cost many of them their lives. The Haitian slave revolution had begun in earnest: "[I]n a few days one-half of the famous North Plain was a flaming ruin. From Le Cap the whole horizon was a wall of fire" (James 1980: 88).

For the next thirteen years, black military units fought against local whites, mulattoes, European soldiers, and each other in groups ranging from scores to thousands. Their weapons ranged from cutlasses to field artillery captured from routed white troops. Their tactics sometimes were frontal assaults on fortified estates, towns, and cities, but against invading troops they employed the classic hit-and-run tactics of guerrilla warfare. A vital element of the revolution was vodun, the Haitian religion that was a syncretism, or combination, of West African belief and European Catholicism. The latter element of vodun allowed the black rebels to absorb and neutralize Christian admonishments against violence. The former element was spiritually the more powerful, involving a recent African heritage and shared survival strategies against plantation cruelties. Boukman was only one of the revolutionary leaders practicing religious beliefs in conjunction with the rebellion. And "it is no accident" that all of the black leaders of the Haitian revolution – including the famous Toussaint Louverture – were "triumphant denizens themselves of the vodun pantheon" (Lewis 1983: 194).

France attempted to maintain control over the colony amidst the turmoil, exporting its own brand of revolutionary change that drove St. Domingue's white planters into the arms of the British. The radical French Republican Commissioner, L. F. Sonthonax, in 1792 began to promote St. Domingue's mulatto officials over local whites, imprisoning many of the latter. Eventually finding his position supported almost exclusively by non-whites, Sonthonax abolished slavery in St. Domingue in August, 1793. The following month, British soldiers invaded the colony. The invasion was part of a British strategy to capitalize on near-anarchy in France to take control of France's Caribbean colonies (and reenslave the blacks in those colonies.) Britain planned to divide St. Domingue with Spain, the Spaniards taking the north and west and the British taking the south and most of the important ports (Ott 1973: 76–77). After a promising start, the British campaign ended in failure, and they withdrew in 1798, having lost 60 percent of the men they had sent there (Geggus 1989: 38–39). Spain also had committed troops to the fray, invading from the east, but the Spaniards pulled out of St. Domingue in mid-1795, ceding Santo Domingo – the eastern half of Hispaniola – to France.

The black armies that defeated the British, and Spanish, and also local black mercenaries hired by the invaders, owed much of their success to the leadership and cunning of the black general Toussaint Louverture. A trusted coachman at the beginning of the revolution, Toussaint rose to become the *de facto* ruler of St. Domingue in 1798 because of his startling military victories over the Europeans. At first allied with Spain, later with France, and always adroitly playing off one local revolutionary group against the other, he epitomized the bewildering complexity of the revolution itself. Toussaint understood brown and white segments of the colony despite his own diminutive stature and very dark skin. He was a tireless worker, familiar with medicinal plants, and reputed to have supernatural powers. Toussaint Louverture was tricked into becoming captured and deported by French authorities, and he died in a dungeon in eastern France in April, 1803.

Disease, probably yellow fever, unquestionably killed more European soldiers coming to St. Domingue than did opposing armies during the revolution. Among the estimated 20,700 British troops sent to the colony, an astounding 12,700 are estimated to have perished there, with perhaps only 1,000 dying in combat. The first British troops arriving late in 1793 enjoyed a fairly healthy winter, but many perished from disease during the following summer. The concentration of soldiers in fortified ports combined with their lack of immunity heightened the spread of yellow fever since its vector, the *Aedes* mosquito, bred in puddles created by human occupation. Malaria also was responsible for some of the deaths as was dysentery; precise diagnoses, of course, are difficult to determine from historical reports. Poor sanitation, the general ill health of troops upon departure from Europe, and excessive alcohol consumption further contributed to the soldiers' high mortality rates, and European commanders complained ceaselessly about their sick troops' inability to match the vigor of local black soldiers (Geggus 1982: 347–372).

The black rebels' final victories, ending in political independence, came against a French reinvasion. Initially welcomed by many black dissidents early in 1802, the French actually intended to impose slavery once again, a ploy that became apparent later in the year. The war in St. Domingue – for over a decade a bewildering series of battles involving blacks, browns, whites, and shifting allegiances – suddenly boiled down to an elemental and bloody black versus white conflict in late 1802. Amidst incredible cruelty on both sides, with yellow fever decimating French soldiers, and with the British bombarding French-held ports, the French armies capitulated in November, 1803. On January 1, 1804, the black soldier-president Jean-Jacques Dessalines – a recent ally of the French and no stranger himself to inflicting cruelty – proclaimed a new republic with the Amerindian name of "Haiti." "I have given the French cannibals blood for blood," announced Dessalines. "I have avenged America" (Geggus 1989: 47).

## Insurgency in the Greater Antilles

Fidel Castro's revolutionary success in Cuba in the 1950s could be interpreted, taking a global view, as indicative of a surge of world Communism at mid-century. For Cubanists, his triumph usually is seen as the result of an external force (Castro's guerrillas) tapping the discontent of the varied Cuban social classes and thereby eventually dominating the island's rival sociopolitical factions. But the Cuban revolution also can be explicated as an extension of a long-smoldering tradition of resistance kept alive in eastern Cuba since the late nineteenth century. Oriente province was Cuba's last frontier, the region's lands finally engulfed by US-financed sugar interests in the late nineteenth and early twentieth century. This engulfment involved seizures of peasant lands, sporadic resistance, and active patrolling of the countryside by US Marines as late as

1922. The spirit of resistance then lived on in eastern Cuba so that when Castro and his dozen or so followers retreated into the Sierra Maestra of southeastern Cuba after their failed uprising in December, 1956, they entered a forested mountain zone of the Cuba countryside that had "a tradition of rebellion, however vague and ill-defined" (Pérez 1989: 192).

In an even broader sense, the torch of slave rebellion may be said to have been passed from Haiti to Cuba one and one-half centuries earlier. The St. Domingue revolution had sent many French planters, who took their slaves, to Cuba, each group thereby importing tales of horror or freedom – depending on whose version was being told – to the larger Spanish island. One Cuban slave, who was in direct contact with Haitian rebels, attempted an islandwide revolt as early as 1810 (Ott 1973: 194).

Fear of a Haitian-style slave rebellion may help to explain why the post-Napoleonic collapse of the Spanish New World Empire did not extend to the Caribbean. The early nineteenth century saw the establishment of Spanish American "settler independences" from Mexico to the Andes, new nations dominated by local economic elites (Wallerstein 1989: 250–256). Spanish Cuban planters, in contrast, were "perpetually haunted" that open rebellion would lead to another Haiti. By the 1830s, however, British anti-slavery pressure on Spain helped lead to a Spanish policy of encouraging white Spaniards to emigrate to Cuba; over 35,000 Spaniards, mainly from the Canary Islands, had emigrated to Cuba by 1839 reversing the idea that only non-whites were fit for plantation work (Knight 1970: 113–115).

Yet Cuban rebellion was not aborted, only delayed. In the following decades, the stresses of rapid economic change in Cuba, marked by the importation of thousands more from Spain as well as thousands of new African slaves, eventually exploded into open warfare in the Cuban countryside from 1868 to 1878. Freedom for slaves was only one of the motives among the Cuban forces opposing Spanish rule in the Ten Years War, and the peace treaty of 1878 again stalled Cuba's political freedom from Spain. During the war, fifty thousand Cubans died, hundreds of sugar mills were destroyed in the central and western parts of Cuba, and the coffee, livestock, and tobacco of Oriente province were ruined (Schwartz 1989: 44). The pitiless economic depression of the 1880s following the Ten Years War had its external origins, of course, in heightened European beet-sugar production and the subsequently lowered world sugar prices (Chapter 3). And conditions in that decade were all the worse in Cuba because of the post-war destruction and a series of new local taxes. Further, tens of thousands of new immigrants from Spain – dirt-poor families from Galicia and Asturias in the north of Spain – flooded the already overcrowded Cuban labor market, their trans-Atlantic voyages subsidized by the mother country starting in 1886 (Pérez 1989: 5–11).

It is not surprising that in these years of Cuban turmoil accented by wrenching social change, new labor regimes brought about by immigration and slave emancipation (1886), and an increasingly heavy-handed Spanish colonialism, a spirit of banditry emerged in the countryside. Gangs of bandits roamed rural areas of Cuba in the 1880s, extracting payoffs from anxious sugar-cane estate owners so that their crops' acreage would be protected. The bandits drew logistical support from the rural populace and usually eluded the police detachments hounding them. The imposition of martial law by Spanish colonial authorities in 1888 and again in 1890 was aimed at tracking bandits down. These lawless bandit groups played active roles in the bloody Cuban war of independence from 1895 to 1898, although present-day historians are divided as to whether the Cuban bandits were motivated by an ideology of political independence and freedom (Pérez 1989) or a less admirable self-indulgence that took personal advantage of the general chaos at the time (Schwartz 1989).

The war in the 1890s devastated the Cuban landscape. By early 1896 the rebel army was estimated at 60,000 armed men; it literally burned its way through the canelands of central Cuba, from east to west. In retaliation, the Spanish general Valeriano Weyler, a veteran of the Cuban wars two decades earlier, adopted a scorched-earth policy; he ordered the human evacuation of strategic areas, then he destroyed the crops, livestock, and dwellings therein to deny them to the rebels. The eventual Cuban victory, which was redirected and blunted by US intervention (Chapter 4), was achieved at the expense of incredible physical destruction and a loss of 10 percent of the island's human population owning to battle casualties, malnutrition, and associated disease (Schwartz 1989: 239–242).

Cuban resistance and revolution in the late nineteenth century took on circum-Caribbean dimensions. After the Ten Years War of 1868–78 a number of Cuban exiles ended up in Hispaniola, Central America, and Florida. Cuban José Martí was arrested during the Ten Years War and then sent to Spain where he studied law. Martí returned to Cuba, then went on to New York where he became a leader of the Cuban emigre community, and he eventually led the

Cuban independence movement in the 1890s. The black Cuban rebel leader Antonio Maceo, killed in combat by the Spaniards in 1896, had plotted insurrection a decade earlier while in Jamaica and had been instrumental in rounding up anti-Spanish recruits in Central America. Other rebel leaders commuted back and forth to Key West, Florida, deriving support from exiled Cuban cigar manufacturers there (Schwartz 1989).

Outbursts of protest in Cuba into the third decade of the twentieth century inspired repeated US intervention despite the formal American withdrawal from the island in 1902. Civil strife after 1905 elections was followed by US military occupation from 1906 to 1909. Then American marines landed in Oriente province in May, 1912, to protect North American holdings – railroads, company stores, livestock – from (mainly black) rioters; possibly as many as 3,000 blacks were killed (Jenks 1928: 115). US Marines returned to Oriente in 1917 and stayed until 1922 involved in "maneuvers" to track down an estimated 10,000 insurgents who burned and looted; from most reports, the insurgents were disenchanted smallholders uprooted by the spread of US sugar-cane holdings (Pérez 1989: 184–88).

The American military intervention in Hispaniola in the second decade of the twentieth century (Chapter 4) elicited tenacious, grassroots guerrilla movements in both Haiti and the Dominican Republic. The conflicts between American military detachments and local irregulars were by no means identical in the two halves of Hispaniola. Nevertheless broad similarities apply to both cases, suggesting something approximating predictability when technically superior armed forces representing an imperialist power confront locally based peasant resistance.

In both places US marines confronted elusive foes employing hit-and-run guerrilla tactics, using ambushes effectively, and dominating the countryside at night when the US soldiers withdrew into encampments or villages. Guerrilla forces in both places derived logistical support – either through sentiment, intimidation, or a combination of the two – from local peoples. Limited aerial bombing and strafing by US forces had little effect in either place except to drive local villagers into closer contact with rebel forces. The countrysides were physically devastated in both Haiti and the Dominican Republic. In both cases frustrated American soldiers, unable to come to grips with the shadowy opponents they categorized as "gooks" or "niggers," committed murders and atrocities. And, in both cases, the marines withdrew without clear-cut victories, underlining the ultimate pointlessness of both military campaigns. It hardly needs emphasis that reading the available accounts of the US military adventures in Hispaniola in the early twentieth century inspires an eery feeling of *déjà vu*, and one wonders if the history of the United States during the 1960s and 1970s would have been substantially different had US State Department officials appreciated fully the lessons that Americans might have learned from Hispaniola decades earlier.

The Haitian rebels were known as *cacos*, a fraternity said to identify themselves to each other with tiny wisps of red-colored thread hidden inside their clothing. *Caco* ranks were most numerous in rural areas in the north. There the United States directed development programs involving roadbuilding that used unwilling laborers who had been conscripted from local villages and set to work under the supervision of US marines and Haitian police. American officials never were certain as to how many *cacos* they were facing; the marines claimed that their enemy consisted of roughly 2,000 "bandits" whereas *caco* leaders claimed a following of 30,000–40,000.

The Haitian rebels fought where and when they wanted to fight, just as they had against European soldiers more than a century earlier. US rifle squads and platoons patrolled the Haitian countryside by day, encountering infrequent ambushes made all the more frightening by *cacos* blowing on conch shells during their attacks. Americans summarily executed Haitians in a few cases, the number of such illegal executions estimated as 400 by one American officer who had supervised road construction crews. The charismatic *caco* leader was Charlemagne Peralte who was murdered by two marines in 1919 but whose inspiration endured thereafter; the Marines photographed his propped-up corpse that had been placed in such a position that it bore a resemblance to Christ on the cross, and when the photo was distributed throughout Haiti with the intention of convincing the Haitians that Peralte was dead, it had much the opposite effect (Schmidt 1971: 100–5).

Haitian riots a decade later led indirectly to eventual American military withdrawal from the country. Late in 1929 a general students' and workers' strike protested the continuing US presence in Haiti, by then an occupation of fifteen years' duration. Early in December in Les Cayes, in the south, about 1,500 "angry peasants" surrounded a marine detachment which then opened fire, killing between twelve and twenty-four rioters, depending on whose estimate was used. A commission from the United States investigated, a hurried program of "Haitianizing" government programs formerly under US military control followed, and the last marines finally withdrew from Port-au-Prince in 1934 (Schmidt 1971: 196–207).

The resistance to the American military occupation of the Dominican Republic derived its muscle from the ranks of displaced peasants in the eastern part of the country whose lands recently had been taken over by large-scale sugar-cane cultivation. Rebels usually were from the *bateys*, the villages owned by sugar-cane companies. The resistance itself, which lasted from 1917 to 1922, quickened noticeably each year at the end of the summer cane harvest and the accompanying onset of seasonal unemployment. As in Haiti, US marine patrols into the Dominican countryside seemed never to pin down their adversaries. In 1918 the Americans adopted the tactic of concentrating residents of rural districts in key villages so that the marines could then scour the countryside for insurgents. This population concentration strategy, which was, of course, quite similar to Spanish tactics in Cuba in the 1890s, was notably unsuccessful, eliciting sharp dissent from rank-and-file Dominicans as well as from sugar-cane estate owners who experienced both crop damages and labor shortages owing to the American policy (Calder 1984: 148–152).

The Dominican insurgents numbered perhaps six hundred hardcore fighting men whose food, shelter, and moral support came from the local peasantry. Unable to claim traditional military victories over people and territory, the American forces contented themselves by reporting the heavy losses inflicted on the enemy. For example, at the end of its two years of hunting Dominican rebels, the marines' Third Provisional Regiment was replaced in the field early in 1919; the unit reported heavy casualties suffered by the rebels and only six marines killed in battle (Calder 1984: 153).

"General" Ramón Natera was the most important Dominican resistance leader, fighting against the marines between 1918 and 1922. Just as Toussaint had done, Natera often changed encampments, switched horses, and purposely avoided routine so that his whereabouts would be known by very few. His campaign, perhaps most importantly, was fought for more than local peasant ends. Natera insisted that he was fighting to end the US occupation of the Dominican Republic. The fighting in the countryside ended in a stalemate in 1922, and the guerrillas agreed to a conditional surrender: "It was a capitulation at least partially predicated on the then impending withdrawal of US forces from the republic" (Calder 1984: 115).

In Puerto Rico, discontent with US colonization of the territory resulted in the founding of the nationalist political party in 1917 with considerable pro-independence sentiment. Workers' unrest in the 1930s – especially among male sugar-cane workers and female needlework laborers – fueled anti-US sentiment among Puerto Ricans from all social classes. In October, 1935, an encounter between police and students at the university in Río Piedras left five killed and forty wounded. Eighteen months later, on Palm Sunday in March, 1937, the nationalists marched in Ponce, in the sugar-cane belt on the southern coast of the island. The police opened fire, killing nineteen and wounding an estimated one hundred. An indirect result of the nationalist uprising, which had been led by Albizu Campos, was to rechannel Puerto Rican political protests and energies. In 1938, Luis Muñoz-Marín founded the Popular Democratic Party in Puerto Rico, and he won the islandwide election in 1940 with the slogan "Bread, Land, and Freedom" (Maldonado-Denis 1972).

Fidel Castro's Cuban insurgency in the 1950s, which finally toppled the corrupt Batista regime early in 1959, self-consciously identified itself with the earlier Cuban wars of independence. After Castro and his vastly outnumbered revolutionaries attempted unsuccessfully to overwhelm the Moncada army barracks in Santiago de Cuba on July 26, 1953 (the centennial year of José Martí's birth), Castro was sentenced to fifteen years' imprisonment. Castro then received a pardon and left Cuba for Mexico in 1955 but not before promising to return to deal forcibly with local Cuban despotism because "no other solution remains but that of '68 and '95." And, in the final stages of Castro's eventual military victory in 1958, the "Antonio Maceo" column of bearded insurgents headed into central Cuba from its base in Oriente province, just as Maceo himself had done six decades earlier (Matthews 1975: 67–68, 107).

Castro's return to Cuba from Mexico was via the *Granma*, a tired fishing yacht designed for fourteen passengers; Castro crowded eighty-two aboard the vessel at Tuxpan in Mexico's Vera Cruz state in late November, 1956. The revolutionaries were mainly disenchanted members of Cuba's middle class, such as Castro himself, an erstwhile lawyer. The *Granma*'s voyage to southeastern Cuba was amidst stormy seas; the vessel ran aground off Cuba on December 2nd, a day on which the Cuban army was alert to invasion because of an uprising in Santiago de Cuba that had been planned to coincide with the *Granma*'s arrival. Most of Castro's force was killed or captured, although Castro himself and a handful of survivors escaped into the mountains of the Sierra Maestra of southeastern Cuba.

During the next two years, as Castro's tiny band of rebels grew and strengthened, their mountain location provided not only a physical sanctuary against lowland invaders but also a kind of geographical mystique. The Sierra Maestra "was the roughest, wildest part of Cuba, mostly untracked tree-and-bush-covered jungle." The tough peasants who lived in the area were charcoal burners and coffee growers, *precaristas* (squatters) who lived on the margins of the law. It was an existence and an identity very different from those of lowland Cuba, where urbanism and sugar estate labor were the norms, and the physical contrasts between mountain (*La Sierra*) and lowland (*El Llano*) underscored the geographical distinctiveness of Casto's revolutionaries, although the majority of lowland Cubans also shared his contempt for Batista's rule (Matthews 1975: 78–79).

Little by little, Castro won over the suspicious peasantry of the mountain zone by treating them fairly and honestly, an instinctive strategy scrupulously followed by Mao Tse-tung decades earlier in the mountain fastnesses of northwestern China. Eventually the peasants joined Castro's ranks and provided important support. The revolutionary–peasant alliance was particularly effective because the latter group had relatives living throughout the Oriente area. It was a process that the Cuban rebels came to call "Dressing the guerrillas in palm leaves" (Guevara 1968: 197).

The Cuban revolution benefited immeasurably from news media exposure in the United States that romanticized Castro's exploits and portrayed him as a Robin Hood resisting Cuba's reigning political evil. In February, 1957, *New York Times* correspondent Herbert Matthews, accompanied by his wife, visited Castro in the Sierra Maestra, thereby publicizing his quest and dispelling the reports of Castro's earlier death that had been issued by both the Cuban government and international news agencies. Two months later a television documentary ("The Story of Cuba's Jungle Fighters") was shown throughout the United States on the Columbia Broadcasting System, in which Castro himself assured Americans that he was not a Communist (Black 1988: 104; Matthews 1975; 82–84.)

Several key events marked the eventual crumbling of the Batista government over its last two years. On March 13, 1957, about eighty young men – students and also political opponents – stormed the Presidential Palace in Havana; most not killed in the assault were executed and tortured, thereby helping to radicalize members of Havana's middle class. Early in 1958 Castro's rebels established a radio station broadcasting to the rest of Cuba from the "Territory of Free Cuba in the Sierra Maestra." In May, Castro's forces won an important battle there against the Cuban army; in striking contrast to the cruelty Batista used against captured rebels, Castro released prisoners, thereby achieving a telling propaganda advantage. The final military assault took nearly everyone by surprise. The Antonio Maceo column advanced into central Cuba and took control of the city of Santa Clara in December, 1958. Santiago de Cuba, the city of Oriente province, also fell, and the twin loss led Batista to flee. As it turned out, the rebels under Castro's command never numbered more than 3,000, and it was not until late in 1958 that the Cuban revolution gained truly massive proportions (Matthews 1975: 122). Fidel Castro's revolutionary success benefited from no shortage of luck and good timing. But much of his appeal to fellow Cubans rested on his personal charm and *caudillismo*, leadership with which Cubans could identify in a personal way (Amaro 1987). Fidel Castro also maintained intimate contact with the Cuban people thereafter, a preoccupation long since forgotten by Batista and a characteristic that served Castro well in not only realizing a revolutionary victory but also defending that victory two years later at the Bay of Pigs (Higgins 1987: 144).

### Riots, trade unions, and political independence in the British Caribbean

Unlike the Greater Antilles, where spacious countrysides and somewhat isolated rural populations nurtured small groups of bandits late in the nineteenth century, black descendants of slaves in the small islands of the British Caribbean inhabited village communities near ongoing estates. Further, they depended on the estates for cash wages and, in the smallest places, subsistence itself. Yet black workers in the British islands were not always docile. Groups of blacks regularly protested, petitioned, and sought to better themselves within the colonial legal systems, and they also created disturbances with sufficient frequency to keep planters and colonial officials on edge. Rebellion and riot in Jamaica in 1865 and in Barbados in 1876, the best known of these disturbances prior to 1880, each led to death and destruction.

The sugar depression of the 1880s and 1890s heightened suffering and also protest through the British islands. Lowered wages, planter bankruptcies, and fewer jobs led to angry working-class protests. The explosive atmosphere was ignited in St. Kitts in February, 1896, when rural workers demanding higher wages descended on the capital town of Basseterre, looting shops and smashing windows. That was only one case among several: Dominica (1893 and 1898), Montserrat (1898), British Guiana (1896 and 1903), Trinidad (1903), and Jamaica (1902) all had similar disturbances, some with loss of life and each calling for intervention by either local police or British marines (Richardson 1983: 106). Each riot could be traced to local problems and local personalities, although overarching issues of economic inequality buttressed by colonial control became ever more obvious with an increased regional frequency of working-class disturbances. A British Royal Commission visited the Caribbean colonies in the first half of 1897 to investigate local socioeconomic conditions. A principal recommendation of the commission was that the region's working classes should have greater access to the land as small-scale proprietors rather than only as wage laborers (Beachey 1957).

World War I increased sugar prices and therefore economic prosperity in the British islands, although it also indirectly intensified black protest. The British West Indies Regiment – white officers and black enlisted personnel – had drawn volunteers from throughout the region. Involved in "pioneer" work in France and combat in the Levant during the war, the more than 15,000 soldiers of the regiment suffered nearly 900 battle casualties, and another 1,000 died of disease. At war's end the unit was relegated to demeaning clean-up duty in Sicily. Black noncommissioned officers of the regiment, fed up with the racism they had suffered during the war, formed the revolutionary Caribbean League while in Sicily. They vowed that blacks should govern themselves upon their return to the British Caribbean, using force if necessary. The British War Office and Colonial Office, both alarmed by the rhetoric (which the sergeants thought had been secret) and the obvious potential that combat-trained personnel had for revolution, had two destroyers standing by in the Caribbean region when the regiment was mustered out early in 1919 (Joseph 1971).

The only place where subsequent rioting was tied to returned soldiers was in British Honduras in July, 1919, but the absence of widespread disturbances elsewhere in the region only slightly diminished the fear that colonial officials had of the returned soldiers who felt they had earned equality by fighting alongside European soldiers. Black West Indian expectations were further heightened by the teachings of Marcus Garvey (Chapter 6). Although Garvey's Universal Negro Improvement Association had its headquarters in New

York, branches of the UNIA emerged throughout the British Caribbean by the end of the second decade of the century. Garvey's teachings were especially influential among returned soldiers and no less among migrating men who felt they had earned respect through their arduous sojourns abroad; Garvey himself had worked for the Americans in the Canal Zone in 1911 and had suffered personally from racism and humiliation. Garvey's inspiration was augmented by that of Jamaican Claude McKay, who became a poet and writer in the United States and England and was an early exponent of the concept of negritude.

Chapters of the UNIA were among many interrelated organizations springing up amidst the ranks of black British West Indians after World War I. Workingmen's associations, fraternal organizations, and even religious groupings on the several islands demonstrated a capability to organize. Perhaps it was inevitable that these organizations would take political directions. By the 1920s, especially among the middle classes of the British Caribbean, there were increasing demands for racial equity in the civil services, for more democracy through relaxation of the voting franchise, and for constitutional reform to give broader powers to black and brown peoples on the islands (Williams 1970: 470).

The pent-up aspirations for more freedom again were energized by riots during the depression decade of the 1930s. On the small British islands, where jobs were scarce in the first place, depression conditions reduced livelihood possibilities. And returning migrants, laid off because of depression elsewhere in the region, intensified local unemployment no longer compensated for by remittances from abroad. Despair inevitably led to violence. An estate workers' dispute on St. Kitts in January, 1935, flared into bloodshed. The St. Kitts riots led to an uneven chain reaction of riots that "raced through the Caribbean like fire on a windy day" (Knight 1978: 179). St. Vincent had similar disturbances in the same year. A coal-carriers' strike in St. Lucia in 1935 was followed by a sugar workers' dispute there in 1937. Sugar-cane workers disputed working conditions in British Guiana in 1935 and again two years later. The oil workers' strike centered at Fyzabad, Trinidad, in 1937, was paralleled by urban riots in Bridgetown, Barbados, in the same year. Sugar workers and dock workers in Jamaica protested in 1937 and in 1938. During the explosive 1930s "[e]very British Governor called for warships, marines and aeroplanes; total casualties in the British [Caribbean] colonies amounted to 29 dead, 115 wounded" (Williams 1970: 473–74).

The riots and resistance of the depression decade stand as socio-political watersheds in the history of the British Caribbean. After the riots, sentiments that had been previously articulated mainly by the brown middle classes concerning greater measures of political freedom became the province of the black masses who now were becoming galvanized into labor unions. With the benefit of unions, strikes and work stoppages by sugar-cane workers or dock laborers were no longer isolated and punishable outbursts by frustrated working men; by the 1940s they were quasi-legal means of obtaining better lives, events coordinated by the workers' own leaders and also by increasingly articulate working-class newspapers. The Industrial Trades Union in Jamaica, the Oilfield Workers Trade Union in Trinidad, the Barbados Workers Union, the St. Kitts-Nevis Trades and Labour Union, and similar unions elsewhere all followed the rescinding of local ordinances that had heretofore banned the formation of unions for purposes of collective bargaining. And local labor leaders such as Norman Manley and Alexander Bustamante in Jamaica, Grantley Adams in Barbados, Eric Williams in Trinidad, Robert Bradshaw in St. Kitts, and Cheddi Jagan and Forbes Burnham in British Guiana, emerged to begin to take control of workers' groups and to form political parties that were all but indistinguishable from the unions themselves.

Labor-oriented political parties came to the fore nearly overnight in each of the islands as property and income qualifications were relaxed and universal suffrage granted. Universal suffrage came to Jamaica in 1944, to Trinidad and Tobago in 1945, and to the smaller places shortly thereafter. These changes were accompanied by constitutional adjustments governing local houses of assembly and legislative councils, as representatives elected by the majority vied with nominated delegates for real decision-making powers (Proctor 1962). Although the political changes at mid-century in the British Caribbean were rapid and momentous compared with the cumulative inertia of three centuries of plantation oppression, it still took public protest in some cases to end *de facto* colonial rule. Protests in St. Kitts in October, 1950, for example, publicized the people's impatience with British unwillingness to consult local elected leaders before assigning governors and administrators to British Caribbean colonies. And riot, protests, and random arson in Grenada early in 1951 eventually led to the election of Eric Gairy to Grenada's Legislative Council, formerly the appointed domain of the island's economic elite (Singham 1968: 158–169).

Local labor leaders/politicians assumed ever-increasing political control of local matters throughout the territories of the British Caribbean during the 1950s. But this significant trend was paralleled by increasing regional sentiment in favor of political confederation among the insular colonies as a solution to the problems that an array of Caribbean mini-states would pose in the post-colonial era. A conference at Montego Bay, Jamaica, in 1947 – including political leaders and senior civil servants from throughout the region – agreed on federation in principal. Political freedom was granted to Jamaica together with the British colonies of the eastern Caribbean in 1958 as they united to become member states of the West Indies Federation. The federated state dissolved in 1962, an issue considered further in the following chapter.

After the dissolution of the West Indies Federation in 1962, Jamaica became an independent political state as did Trinidad and Tobago. British Guiana became independent Guyana in May, 1966, and Barbados gained its independence that November. Most of the smaller territories then became "Associated States" with Britain, a temporary arrangement whereby each controlled its own internal affairs with Britain in charge of foreign matters. Most of the former British Caribbean colonies – whether Associated States, former members of the federation, or neither – have since gained full political independence: The Bahamas in 1973, Grenada in 1974, Dominica in 1978, St. Vincent and the Grenadines in 1979, St. Lucia in 1979, Antigua and Barbuda in 1981, Belize in 1981, and St. Kitts-Nevis in 1983.

Political independence for the states of the former British Caribbean has not resulted directly from military struggles featuring full-blown battle campaigns with armies of downtrodden peasants eventually vanquishing European troops. Quite the opposite: independence ceremonies in the former British colonies usually have been marked by handshakes, band concerts, and celebrations. Yet older residents of the Commonwealth Caribbean recall with pride the resistance to colonial policies exhibited by protest leaders in the 1930s and their uncompromising demands in the years thereafter which yielded political and social concessions by British policy makers and which thereby influenced directly the nature and timing of political independence itself. Resistance to colonial rule in the British Caribbean occurred for three centuries, beginning with the Maroons. After slave emancipation, sporadic protests by working peoples were increasingly better organized. During the 1930s young leaders such as Uriah Butler in Trinidad and Clement Payne in Barbados led strikes and disturbances, speaking out against local conditions in particular and colonization in general. Schoolchildren in the Commonwealth Caribbean today learn about Butler, Payne, Robert Bradshaw, Eric Gairy, and the others who led resistance against colonial rule. In their own way, those leaders of the peoples of the former British Caribbean provided inspiration to resist external control as did the great revolutionary leaders of the Greater Antilles – from the early Arawak chieftains to Cudjoe, Toussaint Louverture, Fidel Castro, and many others – over a span of five centuries.

# 8

## Towards a geography of Caribbean nationhood

At the end of the twentieth century, and also at the end of a half millennium of external political control, most Caribbean societies have achieved political independence. The principal exceptions are Puerto Rico and the US Virgin Islands, the French overseas departments, and a small number of tiny islands that continue to maintain formal links with Britain and the Netherlands. All Caribbean societies, whether formally independent or not, retain strong relationships with their previous colonial rulers as well as complex ties with the outside world in general through the necessity to trade and migrate. And all Caribbean societies, on the basis of location alone, remain geopolitically dominated by the United States.

The region's fragmented insularity, combined with its relative proximity to the North Atlantic metropoles, helps to explain why the Caribbean has been so intensely colonized and recolonized over the centuries. Beginning in the late fifteenth century "the Caribbean became a center of activity as direct sea mobility . . . was the key to empire and riches" (Maingot 1989: 259). In the centuries thereafter the region's direct accessibility to ocean transport and travel and the closeness of all Caribbean locales to the sea together suggest that a certain geographical inevitability may have helped to predetermine the Caribbean's long-standing colonial role in history. Put another way, "it is vital to remember that . . . all of the Caribbean territories [have] from the beginning been forcibly incorporated into the world economic system, and particularly that system in the form of Western capitalist core economies. That has meant a system of structural dependency" (Lewis 1987: 149).

Incorporation, however, is not homogenization. The same insular fragmentation that has provided easy access also signifies discrete

184

island realms that have in turn defied simplistic geopolitical lumping together, regardless of how hard outsiders have tried. In 1823, John Quincy Adams, soon to become the sixth president of the United States, proclaimed that "laws of political as well as physical gravitation" ultimately would mean that Cuba (and presumably other Caribbean islands) eventually would "gravitate . . . toward the North American Union" (Williams 1970: 410). Adams was not the first, nor would he be the last, outside observer to stand over a map of the Caribbean and proclaim inter-island groupings or affinities in the abstract, where few existed on the ground. And, as discussed further on, insularity continues to bedevil plans for confederation and groupings throughout the region today as it has before.

A corollary of the Caribbean's geopolitical fragmentation is the small size of most of its political states, and smallness itself presents distinctive political characteristics and difficulties (e.g., Benedict 1967; Clarke and Payne 1987). It strikes most North Americans and Europeans as odd or even farcical, for example, to learn that any prolonged residence in one of the smaller Caribbean states often is marked with personal encounters with the head of state, perhaps at formal receptions or just as often at grocery stores or restaurants. This kind of sustained personal contact between ordinary people and leading politicians, unheard of in the US or UK, is but one dimension of what has been called "exaggerated personalism" in the political make-up of small states in the Caribbean and elsewhere (Sutton 1987: 15–17). As an outgrowth of smallness, personality and charisma play vital roles in Caribbean politics, far outweighing party, platform, or issues in most cases. Although personal political domination is most noticeable in the tiny places, it also is entirely applicable in the Greater Antilles: The Duvalier family, only recently ousted from power, ruled Haiti for three decades; Jamaican politics boil down to a contest between Michael Manley and Edward Seaga, a personal/political rivalry that has been ongoing for more than a decade; a similar rivalry between Joaquin Balaguer and Juan Bosch has dominated Dominican politics for a quarter of a century; and Cuba, the largest of all insular Caribbean states, has been controlled for more than three decades by the *líder máximo* Fidel Castro, doubtlessly the region's preeminent political personality of the century.

Smallness and personalism are related in turn to an inevitable pervasiveness of Caribbean governments over the land they control, an omnipresence that seems inversely related to the size of insular states (Lowenthal 1987b: 43). On the smaller islands, local governments control or have major roles to play in nearly all social and economic enterprises, from cane harvesting to roadbuilding to the allocation of land, commercial licenses, and jobs of every description. Nepotism based on kinship or political alliance is inevitable under these conditions. Caribbean peoples, further, seem generally to accept the all-powerful role that "government" plays, not through fatalistic acquiescence but as a commonly understood part of daily life. The omnipresence of small-island governments is usually confined to single islands; *de facto* political clout does not travel well over water in multi-island states. The residents of smaller, outlying islands almost invariably feel neglected and even exploited by politicians from the larger places.

Small Caribbean states, like mini-states everywhere, are considered comically inadequate by metropolitan journalists (all from large states) who create and color international perceptions. These peceptions suddenly turn sour if menacing impropriety, international crime, or political radicalism lead to tensions that threaten to disrupt regional political equilibrium. In such cases, smallness is no longer humorous but representative of a political power vacuum with dark implications. And the geopolitical history of the Caribbean region in the twentieth century is accented by a series of military maneuvers – usually, but not exclusively, US maneuvers – that periodically fill these vacuums and restore the equilibrium prescribed by North Atlantic metropolitan states. Grenada in October, 1983, and Panama in December, 1989, are only the most recent examples.

External perceptions of what Caribbean politics ought to be are a part of the region's fate over the centuries. Whether they be concerned with landscapes, economic activities, or ideological stances, outsiders always have influenced and often have decreed the directions local peoples should take. Yet Caribbean peoples, in adapting to external influences – whether they be concerned with Black Power or Protestant evangelism – always have modified rather than mimicked, often creating something entirely new or different. This frustrating characteristic of Caribbean peoples to go their own way and to resist outsiders' attempts "to fit people to theory" (Naipaul 1984: 70) has disappointed capitalists and Communists alike for decades. And there is little reason to suspect that these peoples will easily be force-fitted into preconceived models of the future.

This baffling tendency of Caribbean peoples to defy the plans that outsiders have concocted for them extends to Caribbean politics and Caribbean identity, not only in choosing their own political directions but also in displaying ambivalences that cannot be captured by either/or categories. For example, this final chapter discusses, among other things, the tenacity of small Caribbean islanders in preferring their own ministates to broader political unions. At the same time, these same small islanders are capable of exhibiting strong regional identities in exulting over a victory of the "national" West Indies cricket team at Lords or watching televised highlights of the Caribbean carnival in Brooklyn each September. And it may be important, in attempting to unravel or analyze these ambiguous political tendencies, to point out that mutual solidarity among Caribbean peoples always seems strongest when facing the outside world.

## Caribbean control of Caribbean lands

In March of 1989, the law courts of Antigua and Barbuda anticipated hearing the opening rounds of a law suit filed by the residents of the smaller member (Barbuda) of the two-island state that received its independence from Britain in 1981. The suit involves the issue of landownership on Barbuda, specifically whether Barbudan land is owned by Barbudans themselves or by the larger state whose capital is in Antigua. The specific issue that has inspired the litigation is recent sand-mining on Barbuda that has been authorized by the government in Antigua. During the late 1980s a sand-mining company has removed a bargeload of pink coral sand each week from a peninsula off the southwestern tip of Barbuda and has then shipped it to Antigua and other nearby islands for construction purposes. Barbudans and their allies assert that Antigua's financial profiteering from the exploitation of Barbudan resources is only one part of a serious overall problem that the sand-mining creates for the smaller island. Much more important is that the reduction of so much sand from the beaches creates the risk that Barbuda's peninsula of Palmetto Point could be washed away in the event of heavy seas and Barbuda's only village, Codrington, could be threatened under similar circumstances (Coram 1989).

The sand-mining operation on Barbuda, insignificant for the region as a whole and virtually unknown – as is Barbuda – to most of the outside world, exemplifies issues that are at the heart of any discussion involving the political destinies of multi-island Caribbean states. Barbudans consider themselves a sober, family-oriented, and proper people, content in their isolation. Their view of most Antiguans, in contrast, is that of a free-spending, improvident lot who have sold themselves to international banking interests. That is much of the reason why Antiguan domination of the two-island state rankles the Barbudans so. Not incidentally, identical stereotypes are maintained throughout the Caribbean among residents of the smaller islands in multi-island states; the few exceptions are when these smaller places have become completely dominated by tourism such as in the Iles des Saintes, nominally a dependency of Guadeloupe but really a French tropical resort.

But the main reason Barbudan sand-mining has touched a common local nerve is because it involves the removal of Caribbean land. A scattered but persistent body of evidence suggests that Caribbean peoples often have sincere, even reverent, attitudes toward their land (Besson and Momsen 1987). Whatever the reasons for these complex and widespread feelings by Caribbean peoples, they modify greatly oversimplifications that accompany conventional agricultural dogma in the region, e.g. that young people in the Caribbean "refuse" agricultural work because of the memory of slavery. The control of land signifies freedom and provides a partial buffer against external economic oscillation. This significance, in turn, helps to explain the jealous protection of local land resources even, as in the case of the Barbuda–Antigua dispute, against fellow West Indians.

As with all else in the Caribbean, perspective about local attitudes toward the land is provided by the region's history. The institution of slavery itself was, in part, a means of coercion vital for colonial planters so as to deny imported Africans access to locally available lands. Yet it may have been during slavery that slaves began to regard the land in their own village areas and provision plots as "a means of defining both time and descent" and to venerate interred ancestors in association with local grounds (Mintz and Price 1976: 34). As noted in the previous chapter, a key objective of runaway slaves in the region was the control of local lands. Then, after slavery, the principal factor helping to explain the varying nature of proto-peasantries throughout the Caribbean, as well as the lengths to which planters could go to coerce former slaves to labor on estates, was the availability of land (Green 1984). The importance of land and associated attitudes toward it were by no means confined to Afro-Caribbean working peoples during slavery and thereafter. If anything, the hundreds of thousands of indentured Asian Indians who followed black slaves into the region regarded land with even greater importance and greater tenacity. And their descendants who constitute the small-scale "Asian" peasantries in Guyana and Trinidad at the end of the twentieth century provide ample evidence of the continuity of land's importance to local working peoples of the region.

These attitudes about land and identity are perhaps most noticeable on the smallest islands such as Barbuda. On St. Vincent, for example, land has been interpreted as the basis not only for subsistence and cash production but also for pride and self-respect (Rubenstein 1987: 174–75). But the desire and struggle for Caribbean lands by Caribbean peoples have been manifest throughout the Greater Antilles as well. In 1896, the leaders of the Cuban insurgency considered land control an expression of political freedom; a comprehensive agrarian reform decree in that year stipulated that all lands conquered or confiscated by revolutionary forces and not used for government purposes "shall be divided among the defenders of the Cuban Republic against Spain" (Pérez 1989: 49). The emergence of the victorious Popular Democratic Party in Puerto Rico early in the 1940s was tied closely to a parceling out of small blocks of land for the peasantry, in reaction to large-scale land ownership (mainly by US companies) on the island (Silvestrini 1989: 155–56). The motto of the Communist Party of the Dominican Republic, *Pan y Tierra* (Bread and Land) is self-explanatory. And, in Jamaica, where so much of the coastal land has been monopolized by plantations and the interior by foreign-owned bauxite-mining companies, the active quest for Jamaican lands by Jamaican peoples dates back to the Maroons. The late Jamaican economist George Beckford captured the idea with succinct terseness that could be applicable for much of the Caribbean region: "The story of the Jamaican people is essentially the story of a struggle for land" (1987: 1)

The viability of Caribbean nationhood or of any political unit is ultimately based upon the unequivocal control over a given land base. This imperative, especially in light of the cumulative historical control by outsiders over the Caribbean region, perhaps explains why most Caribbean politicians are so repetitively adamant about controlling their own land and resources. Even though the quality of much of the region's land has been reduced owing to colonial exploitation, it remains a vital basis for any success political autonomy may have in the region. And the success that local Caribbean governments have in buffering attempted outsiders' control of Caribbean lands may have much to do with the region's future direction.

It is hardly a revelation, of course, to suggest that "land for the people" is a crucial theme for any newly independent Third World country. And the quest for land may be said to have been at the heart of the Caribbean's most important social revolution, the Cuban revolution of the late 1950s. Fidel Castro signed the country's Agrarian Reform Law on May 17, 1959, which led to the expropriation of thousands of estates, ranches, farms, cane mills, and livestock centers. According to the official Cuban newspaper, the revolutionary land reform scheme "broke the backbone of power of the exploiting classes and imperialism" in Cuba (Matthews 1975: 154). Most of the Cuban land was then consolidated into large, state-run farms on the Soviet model, although roughly 100,000 smallholders received plots with a "vital minimum" of 67 acres each. A second Cuban Agrarian Reform Law followed in 1963. By the mid-1980s, roughly one-fifth of the agricultural land in Cuba was in private hands, although individual farmers were encouraged by the government to work their lands communally. Collections of farms were grouped into cooperatives that cultivated both subsistence crops and export staples, and the Cuban government provided agricultural inputs, such as seed and fertilizers, and also farming advice (Boswell 1989: 131–32).

Although local control of local lands and local resources is crucial for the validation of Caribbean political autonomy and economic vitality, it does not follow that excessive government control, however well-meaning, will be successful. Recent experiences in Jamaica and Guyana, moreover, suggest that much the opposite may be true. From 1972 to 1980 Prime Minister Michael Manley of Jamaica attempted to locate and pursue a so-called "third path" of development, a route lying somewhere between the Cuban and Puerto Rican models that would encourage greater Jamaican control of the local economy while not discouraging outside investment. Manley's policies involved action on several fronts, including state control of the Jamaican bauxite industry, developing agricultural cooperatives, and reorienting educational curricula in the primary and secondary schools. Manley's attempts, it is generally acknowledged, were not successful because of maladministration and excessive government bureaucracy. Guyana's misfortunes have been even worse. Guyana declared itself a "cooperative socialist republic" in 1970 and proceeded to nationalize most foreign-controlled businesses and lands in the country. Seized assets and lands then were reorganized into national government-controlled corporations. These economic changes were effected within the context of a highly politicized atmosphere in which the ruling political party, the Peoples' National Congress (PNC), asserted control over all sectors of the Guyanese society with catastrophic economic misery – marked by food shortages and power outages – the result (Thomas 1988: 210–231; 255–65).

Land in the former British colonies of the eastern Caribbean has continuously come under local control since the Royal Commission of 1897 recommended peasant-sized parcels, and nearly a century later land in the small islands is the basis not only for large-scale cash cropping but also for subsistence (Rojas 1984). The sugar-cane industry survives in a number of places in the eastern Caribbean and under varied means of local control: private estates on Barbados with wages mediated by the government; state-controlled plantations on St. Kitts; and medium-sized plots in Trinidad cultivated by cane farmers of East Indian descent. The Windward Islands' banana industries, as discussed earlier, thrive with local growers cultivating plots that are privately owned and also rented; their output is focused toward the protected British market which is of essential importance. The agricultural mainstay in the region is foodcrop farming – for home use and increasingly for cash – which is "a syncretic adaptation worked out [originally] by African slaves who incorporated African and European elements into aboriginal systems that already existed in the islands" (Berleant-Schiller and Pulsipher 1986: 20).

The wide variety of land uses in the eastern Caribbean, all under various types of local control, is itself a statement about the kind of adaptability that newly independent governments in the region as a whole will have to develop. And political success probably will depend upon walking a tightrope between shrill nationalism and control that will provoke metropolitan reaction and deferential acquiescence to outsiders that denies self-respect. This balancing act will be nowhere more noticeable than in the land-use policies of Caribbean states. These policies must acknowledge the long-standing Caribbean attitudes toward Caribbean lands, attempting to harness these attitudes with an eye to maximizing food production. At the same time, future policies must not romanticize land's subsistence value but should acknowledge fully that rising expectations influenced by external stimuli often create economic activities and a wage-seeking people that "neither have nor (eventually) want land" (Mintz 1964: xxxvii).

## Overcoming insularity: the regional vision

Early in the 1990s, the small states of the Commonwealth Caribbean face at least three crucial negotiations concerning the changing shape of the international market. The results of these negotiations will weigh heavily in determining the region's economic future. First, bargaining will take place during 1990 for a so-called Lomé IV agreement to replace the current Lomé III arrangement which outlines the formal economic rules governing commodity flows from former colonies to certain European countries, including the bananas from the Caribbean's Windwards to Great Britain. Second, the "Uruguay Round" of negotiations in the global General Agreement on Tariffs and Trade (GATT) will be completed with the possible erosion of trade preferences for the African, Caribbean, and Pacific (ACP) states. Third, "1992" in Western Europe looms ever closer, and it is compounded by the dramatic changes and hopes for unification between East and West; all of these changes heighten economic uncertainty in the small Caribbean states that already have viewed 1992 with caution bordering on trepidation (Bryan 1989: 5).

It is probable that tiny Caribbean states will fare better in these and similar negotiations as a group rather than separately. Few small countries can afford financially to support teams of negotiators. A single team representing the entire region also would show a united front and lessen the chance for special arrangements made by each island state which might lead to self-destructive intraregional competition. Unity among these small states, furthermore, would seem to be a sensible strategy in dealing with the world at large. The small states, especially of the former British Caribbean, have much in common locationally, ethnically, linguistically, historically, politically, and economically. Their peoples have traveled back and forth and interacted with each other for many decades. So it seems to make good sense to formalize these natural affinities, and eventually to include non-English speaking Caribbean states, so that all of these countries and their Pan-Caribbean citizenry might prosper in the years to come: "According to the well-known law of international relations, the more friends you have the safer you are. There is safety in numbers" (Lewis 1987: 157). Yet, despite the seemingly obvious advantages of small Caribbean states grouping together, lasting federations have thus far eluded them. Big is not always beautiful in the eyes of Caribbean politicians and Caribbean peoples, and so far the appeal of local sovereignty has outweighed the theoretical promise of inter-island unification.

The refrain of Caribbean integration is as old as the region's colonization. John Quincy Adams and some other nineteenth-century American observers, for example, foresaw the region united under the American flag. The British Colonial Office grouped and labeled its Caribbean possessions in a number of related ways through history – as multi-island colonies, presidencies, Associated States – in various arrangements that often made more sense to record-keepers in London than to Caribbean residents themselves. And the prospects of various kinds of political, economic, or trade federations in the region are tirelessly reported in newspapers and magazines today as possible solutions to smallness and insularity.

Centrifugal political tendencies in the region usually have out-weighed the appeal or the reality of possible regional political integration. And, taking a long view, physical fragmentation is only one obstacle to such integration. Probably more important is the political shatter zone that is the cumulative result of five centuries of external colonization: the aboriginal peoples of the Caribbean seem to have developed a certain regionwide cultural and political unity, and political balkanization actually was an import from Europe (Lewis 1983: 221–22). A less esoteric and more modern dilemma to some sort of pan-Caribbean integration is that late twentieth-century schemes dealing with possible political groupings in the region seem to be the principal preserve of English-speaking West Indians; representatives of the larger and more populated Spanish-speaking islands seem indifferent toward the idea. And Caribbean "integra-tion" with Cuba as an active or leading force implies much more, of course, than a group of small islands trying to prosper together in a changing world.

Yet persistent suggestions from within the region propose varying groupings among Caribbean polities each year, even in the face of dashed hopes and last year's failures. These hopes, especially in light of the remarkable political changes taking place in Europe late in the twentieth century, should be enough to quiet cynics who claim that Caribbean unification is impossible. And hopes for integration in the region actually extend well beyond the English-speaking islands: Puerto Rico, through an extension of its special relationship with the United States, could become a finance and manufacturing center in the region with outliers among the islets of the Common-wealth Caribbean. Even in the French Antilles, tied directly to Paris by departmental status, some sentiment suggests that closer links might be sought with each other (and not simply each tied to metropolitan France) and also that the French islands might benefit from "closer cooperation with other neighboring countries in the Caribbean" (Crusol 1986: 194). Finally, there is ongoing discussion about the possibility of the United States–Canada free trade pact extending throughout the Caribbean and the rest of the Americas, an arrangement that would formalize American economic domination over the Western Hemisphere.

The demise of the West Indies Federation early in the 1960s has provided a sobering object lesson for three decades as to the region's endemic centrifugal forces that can rupture the best of political integration intentions. In 1958, Great Britain organized the West Indies Federation among the majority of its former Caribbean colonies – the Turks and Caicos Islands, the Caymans, Jamaica, the Leewards, the Windwards, and Trinidad and Tobago – in part to avoid granting full independence to a series of seemingly non-viable micro-states. Plans for the federation were developed in consultation with local political leaders and in the heady atmosphere of imminent political independence. But the federation itself seemed doomed at the start. Inter-island acrimony clouded the decision about siting the federation's capital; the Jamaicans and Trinidadians, residents of the federation's two largest territories, expected that it should be located in Jamaica or Trinidad (depending on which group was consulted), and residents of each of the smaller places considered their own island an ideal compromise (Lowenthal 1958). The two larger territories also jealously guarded their own economies and resources against the smaller territories who demanded a free flow of people among all units of the federation. The federation's cabinet, eventually seated in Trinidad, never was effective and thus belittled as "a lot of Federal Ministers running about Port-of-Spain spreading joy." In September, 1961, the Jamaican populace voted against further membership in the federation and in favor of independence alone. Trinidad and Tobago followed suit shortly thereafter, and the West Indies Federation was dissolved formally in 1962 (Payne 1980: 18–19).

Most of the smaller units of the erstwhile federation then reverted to pre-independence Associated Statehood with Britain but further fragmentation threatened these semi-autonomous units that had more than one island. The Anguilla affair was more than a threat and the ultimate in vociferous wranglings among peoples from tiny Caribbean islands. In 1967 Great Britain granted Associated Statehood to the three-island unit of St. Kitts, Nevis, and Anguilla. The latter island, with a population of only 6,000, far removed from the other two, and without sympathy or identity with the St. Kitts trade union government, soon "rebelled" against St. Kitts. Amidst dark rumors that international gangsters were about to descend on Anguilla to take advantage of near-anarchy there, Britain responded by dispatching a crack paratroop battalion to Anguilla, an "invasion" delighting the Anguillians (Westlake 1972). Members of the international news media likewise were delighted to report on what they considered an event of comic-opera proportions, and *Time*'s article of March 28, 1969 ("Britain's Bay of Piglets"), was perhaps the winner in an unannounced competition among journalists for the most humorous article title describing the event. But political leaders throughout the region were far from amused, denouncing what they considered British recolonization. Anguilla, incidentally, became a British Crown Colony once more in 1980, largely because the Anguillians asked to become a British colony again.

The hostility exhibited by residents of very small islands such as Anguilla and Barbuda toward domineering larger political partners is simply the outward face of a strong sense of local community that, in most cases, binds members of tiny places together in mutual solidarity (Sutton 1987: 17–18). Differences among individuals and families in these tiny places often are ignored, and a confining sense of geographic limits perhaps reduces interpersonal friction. Small island residents simply know that they must get along. So there is a related wariness toward any outsiders, even fellow West Indians, whose influence may lead to an unraveling of the fabric of small-place solidarity. These feelings seem evident even in relatively large places in the region. In French Guiana, for example, some fear excessive contacts even with Guadeloupe and Martinique, in part because the tiny South American country "is frightened it will lose its identity if it has too many contacts with its neighbors" (Schwarz-beck 1986: 183).

The glaring exceptions to any anticipated rule that persons among small contiguous land units of the Caribbean are inherently compatible are provided by the multi-ethnic states of the southern part of the region. In Trinidad, Guyana, and Suriname the ethnic rivalry initially created by colonial planters (who flooded post-emancipation labor markets with indentured Asians) continues and, if anything, has been heightened during the era of political independence. And differences among the rival groups in these newly independent states further complicate the difficulties for any prospects of future regional federation. In Trinidad, for example, inter-group hostilities between the descendants of West African slaves and indentured workers from India are mirrored in personal stereotypes, social networks, economic prospects, and political affiliations that pose an underlying threat of upheaval in an economy already severely buffeted by international market forces (e.g., Clarke 1986).

But it is in the mainland countries where ethnic pluralism has underlain violence and political rupture, mainly between Afro-Caribbean and Indo-Caribbean peoples. The colonial economy of British Guiana evolved along well-defined ethnic/ecological lines: most Afro-Guianese sought wage work, often in the few urban areas, and as government and service personnel; the descendants of indentured Indians, in contrast, usually were small-scale rice cultivators, residing in village settlements along the coastal plain (Despres 1967). Representatives of the two groups competed for jobs on the coastal sugar estates and the upriver bauxite mines. The colony gained internal antonomy in 1951, but the new constitution for British Guiana was rescinded by Britain in 1953 because it was feared that Cheddi Jagan – the leading politician and leader of the numerically dominant Indian element – planned to transform British Guiana into a Communist state.

In the late 1950s, local politics in British Guiana split sharply along racial lines, and strikes, work stoppages, and violence flared between the two groups. Jagan led the rural, Indian-dominated Peoples' Progressive Party (PPP) while Forbes Burnham, a black lawyer trained in London and a former political ally of Jagan, emerged as the leader of the urban, black People's National Congress (PNC). Burnham won the CIA-tainted election of 1964 (Chapter 4), and he and the PNC led British Guiana to become independent Guyana in 1966. Guyana's dismal first quarter-century of political independence has seen the country essentially become a one-party (PNC) state and the economy plunge to desperate levels owing to local mismanagement and fickle external markets. Ethnic hatred animates what little is left of Guyanese political life. The PNC has perpetuated itself through well-documented political fraud, and its muscle is derived from a black-dominated Guyanese military force. Guyanese of Indian descent have emigrated by the thousands, seeking refuge in Britain, North America, and a lesser number in Trinidad, many forced to leave assets and savings behind. Alienated Indo-Guyanese continue to dominate the countryside and have even been the targets of occasional (government-backed?) attacks by gangs of blacks, events that have aroused the attention of human rights groups in the United States and elsewhere (Singh 1988: 86–87).

In neighboring Suriname, roughly similar political chaos also has been a distressing feature of political independence. As in Guyana, Suriname's multi-ethnic character is a reflection of earlier colonial labor procurement policies. Suriname possessed an even wider colonial ethnic mix than Guyana (Lowenthal 1960a). In coastal

Suriname descendants of indentured laborers from India (locally called "Hindustanis") and also Javanese descendants inhabited rural villages whereas the capital city of Paramaribo and the upriver bauxite mines were dominated by Surinamese of African descent ("Creoles"). Political rivalries became divided along ethnic lines not unlike Guyana's and led to an astounding emigration stream from Suriname in the mid-1970s; prior to the country's independence in November, 1975, an estimated one-third of Suriname's 400,000 people – mainly those of Asian descent – fled to the Netherlands, fearing the local consequences of a black-dominated government. Suriname's political picture has continued unsettled since independence. The so-called "Sergeants' Coup" in February, 1980, was led by Desi Bouterse and welcomed by perhaps the majority of those remaining in Suriname. But a series of politically inspired and unsettling strikes have since reduced hope for economic stability. Also, a leader of Suriname's military forces was convicted of cocaine smuggling in Miami in 1986, and a sporadic but ineradicable guerrilla movement has raised havoc along the border with French Guiana for several years (Chin and Buddingh' 1987; Brana-Shute 1987).

Yet despite the drawback of physical fragmentation within multi-island states and the inter-ethnic strife in the southern Caribbean, a yearning for some sort of collective identity remains, especially among the smaller states of the Commonwealth Caribbean. That idealized goal has its principal institutional expression late in the twentieth century in CARICOM (Caribbean Community and Common Market). CARICOM, established in 1973, replaced a similar organization, CARIFTA (Caribbean Free Trade Association) that had existed since 1968. CARICOM'S members are the former colonies of the British Caribbean, although Suriname and Haiti earlier have applied for membership. The association's main function is to provide free trade and also tariff protection among members for goods produced in the region, and its long-term goal is to reduce external dependence through regional integration and economic development. The financial arm of CARICOM is the Caribbean Development Bank (CDB). All CARICOM member states have access to the CDB, the latter's external members being Canada, Colombia, France, the United Kingdom, the United States, and Venezuela. Although CARICOM is the principal interstate economic alliance in the Caribbean region, the smaller states of the eastern Caribbean established their own collective organization, the Organization of Eastern Caribbean States (OECS) in 1981 (Maingot 1989).

CARICOM, despite its lofty goals, has not been an economic

success (Ramsaran 1989: 172–78). Many of its problems involve local cynicism toward what people have already heard for years. Various conclaves, summit meetings, and conferences, for example, produce endless rhetoric and newspaper headlines but few tangible results. Difficulty also has accompanied the establishment of tariff rules between what is and is not locally produced. A number of items have been considered "local" only because some of their components have been produced locally. And these loose regulations actually have promoted the kinds of "finishing touch industries" that reinforce external economic control rather than true economic development (Barry *et al.* 1984: 62–63).

Critics of CARICOM suggest that the organization actually is worse than having no organization at all because it deludes Caribbean peoples into a false sense of integration. These critics, further, propose that a more forceful policy program is necessary to unite the disparate elements of CARICOM and the centrifugal tendencies among the member states. Production cooperation, industrial policy, and regional planning would then be coordinated better among CARICOM's member states (Gonzales 1984: 8).

More stringent guidelines and greater all-round control, however, among CARICOM members might necessitate an ideological homogeneity that has not yet existed throughout the region. Although the principal economic thrust among the member states of CARICOM has been capitalist in nature, some of the region's political and economic leaders have had a long-standing affinity with British socialism. And the Commonwealth Caribbean has also developed strong political leaders whose views have been those of a non-aligned, anti-imperialist nature rather than anything else. Prior to their split in the late 1950s, for example, Cheddi Jagan and Forbes Burnham in British Guiana worked together in an anti-imperialist party that entertained a diversity of views (Smith 1962: 178–79). Michael Manley's prescription for a sensible Jamaican foreign policy in the mid-1970s, further, "involving a positive commitment to Caribbean economic regionalism; the search for a common Third World economic strategy; support for the United Nations . . . [and] . . . a commitment to international morality as distinct from cold assessments of self-interest" (Manley 1975: 146) can hardly be called a leftist diatribe. Accordingly, a generally agreed-upon policy of entertaining diverse ideological views within CARICOM was one of the group's original mutual understandings.

The United States, on the other hand, has not interpreted the diversity of approaches to local Caribbean development as merely reflecting a spectrum of ideologies. Under the presidency of Ronald Reagan early in the 1980s, the US saw states within the "Caribbean Basin" (including those in CARICOM) pursuing development strategies that reflected, more narrowly, either Soviet–Cuban Communism or US-backed capitalism. So, the Caribbean emerged in Washington as a "strategic area" (Axline 1988b: 15). As part of its campaign to reduce the spread of what it saw as leftist governments in the region, the US used its financial power by stipulating as a requirement in a proposed loan to the Caribbean Development Bank that Grenada not be given access to the funds. The bank's local directors, several of whom were hostile to Grenada, voted to reject the grant because the bank's charter prohibited it from interfering in the internal politics of any member state: "the whole saga was a vivid illustration of what happens when external definitions and perceptions are brought to bear upon regional aspirations" (Payne 1985: 216). Then the US invasion of Grenada in October, 1983, supported by token forces from the small states among the Organization of Eastern Caribbean States (and thereby also members of CARICOM) created wide divisions within the region. Those on the right, notably Barbados's Tom Adams, Dominica's Eugenia Charles, and Jamaica's Edward Seaga, were accused of conniving to keep the information from those considered hostile to the possibility of invasion, including George Chambers of Trinidad and Forbes Burnham of Guyana. What fragile unity there may have been among CARICOM members prior to the invasion was thereby damaged severely by external pressures.

CARICOM survives into the 1990s as a tarnished hope for unity among the small states of the region. British political scientist Anthony Payne describes the organization's current status with gloomy metaphorical ambivalence: "[T]he patient is alive and likely to survive, but is not expected to recover fully or quickly ... CARICOM is now simply a fact of Caribbean political and economic life which nobody seems to want to destroy but nobody seems able to rescue" (1985: 228). Even in their attempts to join together in order to present a united front to the outside world, small states of the Caribbean have thus been divided by the same kinds of external forces that seem always to have defined the geopolitics of the region.

### Coping with the wider world

Although Caribbean states at times seem preoccupied with seeking some sort of collective political or economic identity, it is certain that a more fundamental concern in each of these small countries is with national security. The hard-won political independence that most of them have achieved in the latter half of the twentieth century is closely guarded against territorial claims, infringement of nearby waters, secession, the use of their territory for illegal purposes and, ultimately, foreign invasion. The military invasion of the Falkland Islands in April, 1982 by Argentina was a chilling reminder to the Caribbean micro-states that such tactics are not a hemispherical prerogative possessed only by the United States to assure right-thinking politics. The Falklands episode further reminded them all that tiny islands everywhere are, by their very nature, vulnerable to the "lusts of outsiders" (Quester 1983).

Yet it is difficult – though not impossible – to envision similarly naked aggression by a non-US combatant in the Caribbean. Such a scenario is unlikely because the twentieth-century American geopolitical domination of the region has become, in the last decades of the century, something approximating total regional hegemony. The political scientist Gordon Lewis asserts, for example, that it was the US invasion of Grenada that signalled the end of a centuries-long loyalty to Britain among the islands of its former Caribbean empire. "Power calls, compels, demands" states Lewis. "More than anything else the significance of Grenada is that it embodied at once the call and the response. For the previous West Indian generation the final accolade of recognition was an invitation to a Buckingham Palace garden party; for the post-Grenada generation it will be an invitation to a social event at the White House" (1987: 145).

The British presence has indeed diminished in the region. In 1990 Britain controls only Anguilla, Montserrat, the British Virgin Islands, the Cayman Islands, and the Turks and Caicos Islands, tiny remnants of its earlier vast Caribbean holdings that once stretched from Belize to Jamaica to Guyana. It appears that Britain will maintain its control over these remaining possessions for the foreseeable future because there is modest support for political independence only on Montserrat. The British also maintain substantial trade and investment in the region, especially with its former colonies from whom it imports sugar, rum, and bananas on a preferential basis. Britain further provides £25–30 million per annum in aid to states of the region, funds geared to improving local infrastructures. Late in the 1980s Britain increased its vigilance against the drug traffic in the region, working closely with the United States. Because of several cases of graft and drug-money laundering in British possessions, London and Washington recently have cooperated to bolster police forces, to gather intelligence, and to attempt to interdict the traffic in general (Payne 1989).

The other former colonial powers also have withdrawn to play relatively minor roles in the region compared with their former presence. France's interests in the Caribbean are, obviously, centered around its three departments of Guadeloupe, Martinique, and French Guiana. Referenda in each of these places after World War II were overwhelmingly supportive of departmental status with France. Since then, Paris has dismissed occasional suggestions about possible political independence for these territories by noting simply that they have been assimilated politically with the metropole; there has been acknowledgement by metropolitan France, furthermore, for improvements in the social and economic status of its Caribbean departments, and development funds have been allocated accordingly. Early in the 1980s pro-independence movements emerged in both Guadeloupe and Martinique based upon economic grievances and deriving their ideology from the thinking of both Frantz Fanon, the Martinique-born psychiatrist who espoused negritude and rejected European domination, and Aimé Césaire, the famous poet and politician from Martinique. One separatist organization claimed responsibility for bombings in Guadeloupe, Martinique, and Paris in 1983 and 1984 that led to the dispatching of French riot police to Guadeloupe. France has attempted to counteract such movements in its Caribbean departments with the institution of locally elected *conseils régionaux* to help in local governance and which provide a greater degree of local antonomy (MacDonald and Gastmann 1984).

The Netherlands has a decidedly waning presence in the region. Unlike the French, the Dutch have attempted to release their Caribbean dependencies in the latter decades of the twentieth century. Suriname's independence in 1975 was not paralleled by the Netherlands Antilles whose far-flung locations (Aruba, Curaçao, and Bonaire off the Venezuela coast and Saba, St. Maarten, and St. Eustatius hundreds of miles away in the northeastern Caribbean) presented different problems. Further, Aruba's "separate status" (and hopes for eventual independence apart from the other Dutch islands) in not wanting to be governed by Curaçao is a particularly vexing issue as the Netherlands attempts to resolve the political destinies of its remaining Caribbean territories. Netherlands Antilleans now have unrestricted entry to the Netherlands, an arrangement that displeases many in the mother country. In 1996, it is planned, some sort of political independence will be granted or forced upon the Dutch islands. Alternatives for their eventual political status range among full independence, Dutch commonwealth status, or the islands together becoming the thirteenth province of the Netherlands (Hoefte and Oostindie 1989).

The resurgence or expansion of Dutch, British, or French colonialism in the region is thus hardly a current Caribbean concern. Yet territorial expansion by geopolitical middleweights at the margins of the region, even under the shadow of the US geopolitical umbrella, is a continuing threat, especially in Belize whose land is claimed by Guatemala and in Guyana, the western two-thirds of which is disputed by neighboring Venezuela. After Belizean independence in 1981, Britain agreed to maintain Belize's external defense (for all practical purposes, defense against the possibility of Guatemalan invasion). And, by the end of the 1980s, with encouragement from the US government, Britain continued to keep in Belize 1,800 troops, a few planes and tanks, and a Royal Navy frigate patrolling the coastal waters of Belize. In all, Great Britain spends about £31 million annually to protect Belize, and Guatemalan, British, and Belizean diplomats hold occasional talks in attempting to arrive at an amicable solution. The British defense outlay is a reluctant expense of former empire, and Britain has been concerned that its military presence in mainland Central America could possibly draw it into wider Central American conflicts (Payne 1989: 24–27).

The territorial claims by Venezuela (which along with the other Latin American countries supported Argentina in its invasion of the Falklands) against Guyana is a continuation of an older Anglo-Venezuelan boundary dispute supposedly settled in 1899. In 1962 Venezuela announced that it would not honor the agreement, an assertion based upon Venezuelan pride and the possible fear of Cuban influence in British Guiana. Shortly after Guyanese independence, the so-called Rupununi uprising in the savanna country of southern Guyana in January, 1969, heightened tensions; an aborted coup involving American ranchers was stopped by the Guyanese army, and the ranchers and their Amerindian supporters fled to neighboring zones of Venezuela and Brazil. Guyana and Venezuela agreed in 1970 on a moratorium to the dispute, but Venezuela refused to extend the moratorium in 1982. Since then the two countries have entered into limited economic and cultural exchanges, most notably a barter arrangement involving Venezuelan oil and Guyanese bauxite. The Guyanese concerns of Venezuelan invasion rose when the US invaded Grenada in 1983 because of Guyana's previous support for Maurice Bishop and Guyana's commitment to socialism. Early in the 1990s the Guyanese fear of its neighbor to the west remains (Singh 1988: 123–30).

In the 1970s and 1980s Venezuela has attempted to promote its own economic cooperation with several Caribbean countries and thereby represents a somewhat recent entry into the region's geopolitics (Axline 1988b). Yet insular Caribbean territories are cautious about Venezuelan motives and her possible territorial and economic interests in the Caribbean. Trinidad and Tobago has had long-standing disputes with Venezuela concerning immigration, fishing in the Gulf of Paria, and also recent drug smuggling (Simmons 1985: 356). Fear of possible Venezuelan annexation of the southern Netherlands Antilles also has lingered for years in Aruba and Curaçao. These two islands were, of course, economically linked to Venezuela's oil industry for decades. Fears of Venezuelan encroachment in Aruba and Curaçao were intensified in the 1970s when the Netherlands announced that it was withdrawing Dutch soldiers from defending the refinery in Aruba and suggested that they be replaced by Venezuelan troops. Venezuela's proximity to the Netherlands Antilles also takes on new meaning as these small islands head towards some sort of political independence.

Cuba has, of course, posed the most serious perceived regional military threat to its Caribbean neighbors in the latter decades of the twentieth century. Although Fidel Castro's explicit aims of exporting guerrilla-style revolution (which were directed principally toward mainland Latin America) were reduced in the 1970s, the Soviet Union has nonetheless supplied Cuba with a formidable conventional military arsenal, in large part to support the Cuban military interventions in Angola and Ethiopia. Cuba now has the most formidable army throughout the broader region of Middle America, including Mexico. Its airforce is the best-equipped in all of Latin America, and its naval capabilities are augmented with Soviet submarines (Payne 1984: 70–71).

The possibility of a direct Cuban military or political threat toward its Caribbean neighbors appears to be receding, most importantly in the minds of other Caribbean political leaders. Cuban socialist revolutionary ambitions within the region perhaps reached their high-water mark with support of the Bishop government in Grenada in the early 1980s. The Cuban involvement in Grenada also followed an intense Cuban presence in Jamaica in the late 1970s marked by intelligence gathering, student and cultural exchanges, the provision of Cuban doctors and nurses to Jamaica, and a strident pro-socialist camaraderie with the Manley government in Kingston.

In 1975 the small states of the eastern Caribbean were incensed when it was discovered that Cuba had made surreptitious use of the Barbados international airport to airlift troops to Angola. And the former British islands of the Caribbean protested vehemently in 1980 when a Cuban jet aircraft attacked and sank a Bahamian gunboat in Bahama's coastal waters. Although Cuba subsequently apologized and paid reparations, nearly 90 percent of Bahamians polled considered the act deliberate (Maingot 1984: 69–71). It is possibly unlikely that similar incidents will animate Cuban activities in the wider Caribbean region considering the momentous changes of 1989 throughout Eastern Europe and quickening rumors about Cuban political changes. Whatever her political stance, however, Cuba will continue to represent a formidable presence in the region possibly outcompeting her smaller neighbors in the future for, of all things, American tourists and even American development funds, providing she reorients her political ideology.

In the early 1990s the small states of the Caribbean are indeed menaced, not by the threats of possible Cuban intervention nor so much by the geopolitical ambitions of other middle-sized Latin American states. The danger to the region's stability lies in the voluminous yet clandestine drug traffic between its South American source and North American market. The Caribbean's intermediate location, as discussed in Chapter 5, underlines its importance as a transit area and also as an offshore financial center for the trade. And the immense sums of liquid capital generated by the drug traffic carry with them the potential of, literally, buying sovereign states. The contemporary drug traffic promises to test the will and resolve of every independent state of the Caribbean region and, as in the past, this particular issue, located geographically in the Caribbean, has global dimensions. The United States and Britain are heavily involved in helping the smaller states of the region control and prevent the flow of drugs; together with Barbados they sponsored a regional drug law-enforcement conference early in 1988 involving police and customs officials from every state in the region (Payne 1989: 22–24). Other outside countries, particularly European states where potential drug profits are even higher than in the United States, are similarly concerned. Dutch concern is growing and with it comes greater surveillance of offshore banking in the Netherlands Antilles. And Martinique already is being described as a "staging post" for cocaine shipments from Colombia to France (Elliott 1988: 16).

## "We're in nobody's backyard"

Facile geographical comparisons, especially when they take the form of slogans, obscure more than they clarify. Some observers, for example, see the Caribbean future possibly modeled upon the economic success of the newly industrializing countries of the Pacific Rim. "Can Jamaica eventually become a Caribbean Taiwan?" asks one observer (Berger 1984: 40). The question is every bit as unhelpful as describing the Caribbean as "The American Mediterranean," an unfortunate musing repeated time and again in textbooks and elsewhere. These comparisons are less than helpful because the Caribbean region – as well as its individual territories – is unique, and its present as well as its future is, and will be, intimately bound up with its particular past.

Unlike Taiwan, or any other place outside the Caribbean for that matter, the Caribbean has been closely linked in a dependency relationship with Europe and North America for five hundred years, a part of capitalism's periphery long before world-economy thinking or the "Third World" ever existed. The Caribbean's indigenous peoples were obliterated almost upon contact. The subsequent importation of massive amounts of labor during centuries of slavery followed by decades of indenture brought a variety of peoples into the region, creating local populations defined by their external origins: "Such movements forged societies of a special sort, in which people became accustomed to jostling with strangers matter-of-factly, accustomed to the presence of different habits, different values, different ways of dressing and of looking, accustomed to anonymity itself, as an expected part of life" (Mintz 1974a: 47).

These imported societies were, however, not "transplanted" African or Asian variants because the initial social and economic activities of these newcomers were invariably mediated by a white plantocracy, itself with origins elsewhere. Upon escape or freedom from plantation confines, Caribbean peoples then faced new environments – often already eroded and always in the shadow of hostile powerholders – without their indigenous customs that had been left far behind. These "native" Caribbean peoples thereby established themselves first and foremost in relationship with the plantation system that had brought them (usually forcibly) into the region in the first place. Their towns, villages and, in many ways, their identities were thus established as responses to the oppression of the local colonial regime. Caribbean lands and peoples, in other words, always were immediate appendages of a larger European-centered economy; there was little indigenous local economy or ecological tradition in which "native" peoples could seek refuge against colonial encroachment (Richardson 1983: 171).

A considerable academic literature thus laments the Caribbean as a colonial concoction, a region where "culture" does not exist in its own right, where ideas and commodities all are imported, and where success or achievement only can be ratified by outsiders. An historically imposed colonialism, according to this line of thought, even creates a psychic malaise among Caribbean peoples: "A too long history of colonialism seems to have crippled Caribbean self-confidence and Caribbean self-reliance, and a vicious circle has been set up: psychological dependence leads to an ever-growing economic and cultural dependence on the outside world. Fragmentation is intensified in the process. And the greater degree of dependence and fragmentation further reduces local self-confidence" (Williams 1970: 502).

A more hopeful, and a more appealing, interpretation of the Caribbean's unique and long-standing relationship with the wider world is that the incongruity between the region's lands and its imported peoples has elicited a "creative nonadaptation" (Patterson 1978: 139). Facing adversity and having to make do with what has been at hand, Caribbean peoples have been forced to survive by their wits and ingenuity. On a mundane level their successful adaptations are mirrored in the great variety of economic activities, including human migration, throughout the region. Some observers, furthermore, suggest that the voluminous outpouring of art and literature from Caribbean peoples is an extension of the remarkable creativity in the region (e.g., Márquez 1989). At a societal level, Caribbean peoples may be said to have equipped themselves with "a whole new set of social inventions" with which to cope with the asymmetrical relationships of "the global system in which they are ... immersed in a mutually dependent way" (Gonzalez 1970: 7).

The era of political independence for the Caribbean region, achieved after five centuries of direct external control, may provide the opportunity to channel this creativity into some sort of structural stability. So it is understandable when newly independent states, even within the American shadow, guard their sovereignty jealously. One month after coming to power in Grenada in 1979, Maurice Bishop addressed the island's populace by radio and, among other things, articulated a now-famous sentiment that could stand for the aspirations of all Caribbean peoples, regardless of their ideological convictions: "We are not in anybody's backyard, and we are definitely not for sale" (Searle 1984: 14). Bishop's remark is an obvious assertion of independence and local pride, but it is also an indirect acknowledgement that Caribbean identity is inevitably framed within the context of the wider world which has played an enduring role in creating the human geography of the region. Whatever direction the Caribbean takes in the future, outsiders' perceptions of it will count for much as they always have before. Yet Europeans, North Americans and others from outside the region need not indulge in condescension, assuming that they must think and act on behalf of Caribbean peoples. Within a global framework, of which Caribbean peoples are quite aware, these people are most capable of seeking their own identities which have so often been denied them in the past.

# Bibliography

Abel, E. 1969. *The missiles of October: The story of the Cuban missile crisis, 1962*. London: Macgibbon & Kee

Adams, F. U. 1914. *Conquest of the tropics: The story of the creative enterprises conducted by the United Fruit Company*. New York: Doubleday, Page & Company

Adamson, A. H. 1972. *Sugar without slaves: The political economy of British Guiana, 1838–1904*. New Haven: Yale University Press

Agnew, J. 1987. *The United States in the world-economy: A regional geography*. Cambridge: Cambridge University Press

Albert, B. and A. Graves (eds.) 1984. *Crisis and change in the international sugar economy, 1860–1914*. Norwich: ISC Press

Alexander, G. 1986. The calypso blues: Why the Caribbean Basin Initiative isn't working. *Policy Review*, 38: 55–59

Amaro, N. R. 1987. Mass and class in the origins of the Cuban revolution. In *Cuban Communism*, ed. I. L. Horowitz, 6th edn. New Brunswick, New Jersey: Transaction Books, 13–36

Anderson, J. 1982. *This was Harlem: A cultural portrait, 1900–1950*. New York: Farrar, Straus, Giroux

Ashdown, P. 1979. *Caribbean history in maps*. London: Longman

Augelli, J. P. 1962. The rimland-mainland concept of culture areas in Middle America. *Annals of the Association of American Geographers*, 52: 119–29

Axline, W. A. 1988a. Political change and US strategic concerns in the Caribbean. *Latin American Research Review*, 23, no. 2: 214–25

1988b. Regional co-operation and national security: External forces in Caribbean integration. *Journal of Common Market Studies*, 27, no. 1: 1–25

Baptiste, F. A. 1988. *War, cooperation and conflict: The European possessions in the Caribbean, 1939–1945*. Westport, Connecticut: Greenwood Press

208

Barrett, W. 1965. Caribbean sugar-production standards in the seventeenth and eighteenth centuries. In *Merchants & scholars: Essays in the history of exploration and trade*, ed. J. Parker. Minneapolis: The University of Minnesota Press

Barry, T., B. Wood, and D. Preusch. 1984. *The other side of paradise: Foreign control in the Caribbean*. New York: Grove Press

Beachey, R. W. 1957. *The British West Indies sugar industry in the late 19th century*. London: Basil Blackwell

Beck, H. 1976. The bubble trade. *Natural History*, 85 December: 38–47

Beckford, G. L. 1972. *Persistent poverty: Underdevelopment in plantation economies of the third world*. New York: Oxford University Press

1987. The social economy of bauxite in the Jamaican man-space. *Social and Economic Studies*, 36 no. 1: 1–55

Benedict, B. (ed.) 1967. *Problems of smaller territories*. London: The Athlone Press

Bennett, H. L. 1989. The challenge to the post-colonial state: A case study of the February revolution in Trinidad. In *The modern Caribbean*, eds. F. W. Knight and C. A. Palmer. Chapel Hill: The University of North Carolina Press, 129–46

Berger, P. L. 1984. Can the Caribbean learn from East Asia? *Caribbean Review*, 13 Spring: 6–9, 40–41

Berleant-Schiller, R. 1977. The social and economic role of cattle in Barbuda. *The Geographical Review*, 67: 299–309

Berleant-Schiller, R. and L. M. Pulsipher. 1986. Subsistence cultivation in the Caribbean. *Nieuwe West-Indische Gids*, 60: 1–40

Besson, J. and J. Momsen (eds.) 1987. *Land and development in the Caribbean*. London: Macmillan

Black, G. 1988. *The good neighbor: How the United States wrote the history of Central America and the Caribbean*. New York: Pantheon Books

Blaut, J. M. 1987. Diffusionism: A uniformitarian critique. *Annals of the Association of American Geographers*, 77: 30–47

Blight, J. G. and D. A. Welch. 1989. *On the brink: Americans and Soviets reexamine the Cuban missile crisis*. New York: Hill & Wang

Bolingbroke, H. 1807. *A voyage to the Demarary*. Norwich: Stevenson and Matchett

Bolland, O. N. 1977. *The formation of a colonial society: Belize, from conquest to Crown Colony*. Baltimore: The Johns Hopkins University Press

1981. Systems of domination after slavery: The control of land and labor in the British West Indies after 1839. *Comparative Studies in Society and History*, 23: 591–619

Bonnet, J. A. and A. Calderón-Cruz. 1985. Caribbean energy dependence: A 15-year prognosis. *Caribbean Review*, 14, Summer: 16–17

Boswell, T. D. 1989. The West Indies: The Hispanic territories and Haiti.

In *Middle America: Its lands and peoples*, 3rd edn. ed. R. C. West and J. P. Augelli. Englewood Cliffs, New Jersey: Prentice Hall, 128–67

Brana-Shute, G. 1987. Suriname surprises: Small country, smaller revolution. *Caribbean Review*, 15, Spring: 4–7, 26–28

Bridenbaugh, C. and R. Bridenbaugh. 1972. *No peace beyond the line: The English in the Caribbean, 1624–1690*. New York: Oxford University Press

Browne, M. W. 1988. Century's fiercest storm sweeps Yucatan. *New York Times*, September 15th

Brubaker, B. 1986. Caribbean curve ball: Baseball's Dominican pipeline. *Washington Post National Weekly Edition*, March 31: 8–9

Bryan, A. T. 1989. The international dynamics of the Commonwealth Caribbean: Challenges and opportunities in the 1990s. *Journal of Interamerican Studies and World Affairs*, 31, no. 3: 1–7

Buckley, R. N. 1979. *Slaves in redcoats: The British West India regiments, 1795–1815*. New Haven: Yale University Press

Burn, W. L. 1937. *Emancipation and apprenticeship in the British West Indies*. London: Jonathan Cape

Butzer, K. W. 1988. Cattle and sheep from Old to New Spain: Historical antecedents. *Annals of the Association of American Geographers*, 78: 29–56

Calder, B. J. 1984. *The impact of intervention: The Dominican Republic during the US occupation of 1916–1924*. Austin: University of Texas Press

*Caribbean Basin Initiative: 1989 Guidebook*. 1988. Washington: US Department of Commerce

Carnegie, C. V. 1983. If you lose the dog, grab the cat. *Natural History*, 92 October: 28, 30–34

Chin, H. E. and H. Buddingh'. 1987. *Surinam: Politics, economics, and society*. London: Frances Pinter

Clarke, C. G. 1983. Dependency and marginality in Kingston, Jamaica. *Journal of Geography*. 82, 5: 227–235

1986. *East Indians in a West Indian town: San Fernando, Trinidad, 1930–70*. London: Allen & Unwin

Clarke C. and T. Payne (eds.) 1987. *Politics, security and development in small states*. London: Allen & Unwin

Clendinnen, I. 1980. Landscape and world view: The survival of Yucatec Maya culture under Spanish conquest. *Comparative Studies in Society and History*, 22: 374–93

Cohen, S. 1988. Jamaican gangs invading American heartland, authorities say. *Roanoke Times & World-News*, January 17

Cohn, M. and M. K. H. Platzer. 1978. *Black men of the sea*. New York: Dodd, Mead & Company

Coke, T. 1811. *A history of the West Indies*. London: Thomas Coke

Collin, R. H. 1990. *Theodore Roosevelt's Caribbean: The Panama Canal, the Monroe Doctrine, and the Latin American context.* Baton Rouge: Louisiana State University Press

Conkling, E. C. 1987. Caribbean Basin Initiative: A regional solution for America's threatened enterprise? *Focus*, 37, no. 2: 2–9

Conniff, M. L. 1985. *Black labor on a white canal: Panama, 1904–1981.* Pittsburgh: University of Pittsburgh Press

Cooper, F. 1980. *From slaves to squatters: Plantation labor and agriculture in Zanzibar and coastal Kenya.* New Haven: Yale University Press

Coram, R. 1989. Ancient rights. *The New Yorker*, February 6: 76–86, 89–94

Cox, E. L. 1984. *Free coloreds in the slave societies of St. Kitts and Grenada.* Knoxville: The University of Tennessee Press

Craton, M. 1982. *Testing the chains: Resistance to slavery in the British West Indies.* Ithaca: Cornell University Press

Craton, M. and J. Walvin. 1970. *A Jamaican plantation: The history of Worthy Park, 1670–1970.* London: W. H. Allen

Cronon, E. D. 1964. *Black Moses: The story of Marcus Garvey and the Universal Negro Improvement Association.* Madison: University of Wisconsin Press

Crosby, A. W. 1972. *The Columbian exchange: Biological and cultural consequences of 1492.* Westport, Connecticut: Greenwood Press

   1986. *Ecological imperialism: The biological expansion of Europe, 900–1900.* London: Cambridge University Press

Cross, M. 1979. *Urbanization and urban growth in the Caribbean: An essay on social change in dependent societies.* London: Cambridge University Press

Crusol, J. 1986. An economic policy for Martinique. *Dual legacies in the contemporary Caribbean: Continuing aspects of British and French dominion,* ed. P. Sutton. London: Frank Cass

Curtin, P. D. 1969. *The Atlantic slave trade: A census.* Madison: The University of Wisconsin Press

Deerr, N. 1949 and 1950. *The history of sugar.* 2 volumes. London: Chapman and Hall

Demas, W. G. 1965. *The economics of development in small countries with special reference to the Caribbean.* Montreal: McGill University Press

Despres, L. A. 1967. *Cultural pluralism and nationalist politics in British Guiana.* Chicago: Rand-McNally

Diederich, B. 1985. Clouds over Aruba. *Caribbean Review.* 14, March: 21

Dirks, R. 1975. Slaves' holiday. *Natural History*, 84, 10: 82–84, 87–88, 90

   1978. Resource fluctuations and competitive transformations in West Indian slave societies. In *Extinction and survival in human populations,* ed. C. D. Laughlin and I. A. Brady. New York: Columbia University Press

1987. *The black saturnalia: Conflict and its ritual expression on British West Indian slave plantations.* Gainesville: University of Florida Press

The dollar impact of St. Croix's troubles. 1973. *Business Week*, August 25: 26

Draper, A. S. 1899. *The rescue of Cuba: An episode in the growth of free government.* Boston: Silver, Burdett and Company

Draper, T. 1971. The Dominican intervention reconsidered. *Political Science Quarterly*, 86: 1–36

Drekonja-Kornat, G. 1984. On the edge of civilization: Paris in the jungle. *Caribbean Review*, 13, Spring: 26–27

Dunn, R. S. 1972. *Sugar and slaves: The rise of the planter class in the English West Indies, 1624–1713.* Chapel Hill: The University of North Carolina Press

Dupuy, A. 1989. *Haiti in the world economy: Class, race, and underdevelopment since 1700.* Boulder, Colorado: Westview Press

Eddy, P., H. Sabogal, and S. Walden. 1988. *The cocaine wars.* New York: Norton

Elliott, M. 1988. Columbus's islands. *The Economist*, August 6: 1–18

Engardio, P., G. DeGeorge, and S. Baker. 1988. The dashed dreams in Gilbert's wake. *Business Week*, October 3: 32

English, P. W. 1984. *World regional geography: A question of place.* New York: John Wiley & Sons

Erisman, H. M. (ed.) 1984. *The Caribbean challenge: The United States policy in a volatile region.* Boulder, Colorado: Westview Press

Eves, C. W. 1893. *The West Indies*, 3rd edn. London: Sampson Low, Marston & Company

Farley, R. 1954. The rise of peasantry in British Guiana. *Social and Economic Studies*, 2: 87–103

Foner, N. 1987. *New immigrants in New York.* New York: Columbia University Press

Forster, N. 1987. Cuban agricultural productivity. In *Cuban Communism*, ed. I. L. Horowitz, 6th edn. New Brunswick, New Jersey: Transaction Books, 196–216

Fortune, S. A. 1984. *Merchants and Jews: The struggle for British West Indian commerce.* Gainesville: University Presses of Florida

Francis, M. J. 1967. The US press and Castro: A study in declining relations. *Journalism Quarterly.* 44: 257–66

Frank, A. D. 1986. New hub for an old web. *Forbes*, 137, April 7: 91–94

Frucht, R. 1967. Caribbean social type: Neither "peasant" nor "proletarian." *Social and Economic Studies*, 16: 295–300

1975. Emancipation and revolt in the West Indies: St. Kitts, 1834. *Science and Society*, 34: 199–214

Galloway, J. H. 1977. The Mediterranean sugar industry. *Geographical Review*, 67:177–94

1989. *The sugar-cane industry: An historical geography from its origins to 1914.* Cambridge: Cambridge University Press

Garthoff, R. L. 1987. *Reflections on the Cuban missile crisis.* Washington: The Brookings Institution

Gaspar, D.B. 1985. *Bondmen & rebels: A study of master-slave relations in Antigua.* Baltimore: The Johns Hopkins University Press

Geertz, C. 1963. *Agricultural involution: The process of ecological change in Indonesia.* Berkeley: University of California Press

Geggus, D. P. 1982. *Slavery, war, and revolution: The British occupation of Saint Domingue 1793–1798.* Oxford: Clarendon Press

1989. The Haitian revolution. In *The modern Caribbean,* eds. F. W. Knight and C. A. Palmer. Chapel Hill: The University of North Carolina Press, 21–50

Genovese, E. D. 1979. *From rebellion to revolution: Afro-American slave revolts in the making of the modern world.* Baton Rouge: Louisiana State University Press

Gilpin, R. 1981. *War and change in world politics.* Cambridge: Cambridge University Press

Girault, C. A. 1985. *El comercio del cafe en Haiti.* Santo Domingo: Taller

Glissant, E. 1989. *Caribbean discourse: Selected essays.* Translated by J. Michael Dash. Charlottesville: University Press of Virginia

Gonsalves, R. E. 1989. *Banana in trouble.* Kingstown, St. Vincent: MNU Educational Pamphlet (mimeographed)

Gonzales, A. P. 1984. The future of CARICOM: Collective self-reliance in decline? *Caribbean Review,* 13, no. 4: 8–11, 40

Gonzalez, N. L. 1970. The neoteric society. *Comparative Studies in Society and History,* 12: 1–13

1988. *Sojourners of the Caribbean: Ethnogenesis and ethnohistory of the Garifuna.* Urbana: University of Illinois Press

Goslinga, C. C. 1985. *The Dutch in the Caribbean and in the Guianas, 1680–1791.* Dover, New Hampshire: Van Gorcum

Green, W. A. 1976. *British slave emancipation: The sugar colonies and the great experiment, 1830–1865.* Oxford: Clarendon Press

1984. The perils of comparative history: Belize and the British sugar colonies after slavery. *Comparative Studies in Society and History,* 26: 112–19

Grove, N. 1981. The seething Caribbean. *The National Geographic Magazine,* 159, February: 244–72

Guerra y Sánchez, R. 1964. *Sugar and society in the Caribbean: An economic history of Cuban agriculture.* New Haven: Yale University Press

Guevara, E. C. 1968. *Reminiscences of the Cuban revolutionary war.* New York: Monthly Review Press

Gugliotta, G. and J. Leen. 1989. *Kings of cocaine: Inside the Medellin cartel.* New York: Simon & Schuster

Hall, D. 1971. *Five of the leewards, 1834–1870: The major problems of the post-emancipation period in Antigua, Barbuda, Montserrat, Nevis, and St. Kitts.* Barbados: Caribbean Universities Press

1978. The flight from the estates reconsidered: The British West Indies, 1838–42. *The Journal of Caribbean History,* 10–11: 7–24

Handler, J. S. 1974. *The unappropriated people: Freedmen in the slave society of Barbados.* Baltimore: The Johns Hopkins University Press

Handler, J. S. and F. W. Lange. 1978. *Plantation slavery in Barbados: An archaeological and historical investigation.* Cambridge: Harvard University Press

Harris, D. R. 1965. *Plants, animals, and man in the Outer Leeward Islands, West Indies.* University of California Publications in Geography

Harvey, D. 1987. The world systems theory trap. *Studies in Comparative International Development,* 22, no. 1: 42–47

Healy, D. 1970. *US expansionism: The imperialist urge in the 1890s.* Madison: University of Wisconsin Press

1976. *Gunboat diplomacy in the Wilson era: The US navy in Haiti, 1915–1916.* Madison: University of Wisconsin Press

Henige, D. P. 1970. *Colonial governors from the fifteenth century to the present.* Madison: The University of Wisconsin Press

Henry, F. 1987. West Indians in Canada: The "victims" of racism? Paper presented at a conference on Caribbean migration. University of London, June 17–19

Henry, P. 1984. *Peripheral capitalism and underdevelopment in Antigua.* New Brunswick, New Jersey: Transaction Books

Heston, T. J. 1987. *Sweet subsidy: The economic and diplomatic effects of the US sugar acts, 1934–1974.* New York: Garland Publishing, Inc.

Higgins, T. 1987. *The perfect failure: Kennedy, Eisenhower, and the CIA at the Bay of Pigs.* New York: W. W. Norton

Higman, B. W. 1976. *Slave population and economy in Jamaica, 1807–1834.* Cambridge: Cambridge University Press

1984. *Slave populations of the British Caribbean, 1807–1834.* Baltimore: The Johns Hopkins University Press

Hill, D. R. 1977. *The impact of migration on the metropolitan and folk society of Carriacou, Grenada.* New York: Anthropological Papers of the American Museum of Natural History

Hill, D. R. and R. Abramson. 1979. West Indian carnival in Brooklyn. *Natural History,* August–September: 72–85

Hobsbawm, E. J. 1967. The crisis of the seventeenth century. In *Crisis in Europe, 1560–1660,* ed. T. Aston. Garden City, New York: Doubleday and Company

1969. *Industry and empire: From 1750 to the present day.* New York: Pelican Books

Hoefte, R. and G. Oostindie. 1989. The Netherlands and the Dutch Caribbean: Dilemmas of decolonization. Paper given at the meetings of the Latin American Studies Association in Miami, Florida, December

Hoetink, H. 1967. *Caribbean race relations: A study of two variants.* Translated by Eva M. Hooykaas. London: Oxford University Press

Howard, R. A. 1973. The vegetation of the Antilles. In *Vegetation and vegetational history of northern Latin America,* ed. A. Graham. Amsterdam: Elsevier, 1–38

Innes, F. C. 1970. The pre-sugar era of European settlement in Barbados. *Journal of Caribbean History,* 1: 1–22

Jagan, C. 1967. *The West on trial: The fight for Guyana's freedom.* New York: International Publishers

James, C. L. R. 1963. *Beyond a boundary.* London: Hutchinson
   1980. *The black jacobins: Toussaint L'Ouverture and the San Domingo revolution.* 3rd edn. London: Allison & Busby

Jenks, L. H. 1928. *Our Cuban colony: A study in sugar.* New York: Vanguard Press

Joseph, C. L. 1971. The British West Indies Regiment, 1914–1918. *Journal of Caribbean History* 2: 94–124

Keller, B. 1989. Warheads were deployed in Cuba in '62, Soviets say. *New York Times,* January 29: 1, 10

Kepner, C. D. and J. H. Soothill. 1935. *The banana empire: A case study of economic imperialism.* New York: The Vanguard Press

Kimber, C. T. 1988. *Martinique revisited: The changing plant geographies of a West Indian island.* College Station: Texas A&M University Press

Kiple, K. F. 1984. *The Caribbean slave: A biological history.* Cambridge: Cambridge University Press

Klein, H. S. 1967. *Slavery in the Americas: A comparative study of Virginia and Cuba.* Chicago: The University of Chicago Press

Knight, F. W. 1970. *Slave society in Cuba during the nineteenth century.* Madison: University of Wisconsin Press
   1978. *The Caribbean: The genesis of a fragmented nationalism.* New York: Oxford University Press
   1985. Jamaican migrants and the Cuban sugar industry, 1900–1934. In *Between slavery and free labor: The Spanish-speaking Caribbean in the nineteenth century.* Baltimore: The Johns Hopkins University Press, 94–114
   1989. Cuba: Politics, economy, and society, 1898–1985. In *The modern Caribbean,* eds. F. W. Knight and C. A. Palmer. Chapel Hill: The University of North Carolina Press, 169–84

Knight, M. M. 1928. *The Americans in Santo Domingo.* New York: Vanguard Press

Knox, P. L. and J. Agnew. 1989. *Geography of the world-economy.* London: Edward Arnold

Laguerre, M. S. 1984. *American odyssey: Haitians in New York City.* Ithaca: Cornell University press

Lang, D. M. and D. M. Carroll. 1966. *Soil and land-use surveys no. 16, St. Kitts and Nevis.* Trinidad: University of the West Indies, Imperial College of Tropical Agricultural

Langley, L. E. 1982. *The United States and the Caribbean in the twentieth century.* Athens: The University of Georgia Press

Lasserre, G. 1978. *La Guadeloupe: Etude géographique.* 3 volumes. Martinique: E. Kolodziej

Latortue, P. R. 1985. Neoslavery in the cane fields: Haitians in the Dominican Republic. *Caribbean Review* 14, Fall: 18–20

LeFranc, E. 1988. Higgerling in Kingston: Enterpreneurs or traditional small-scale operators? *Caribbean Review,* 13 Spring: 15–17, 35

Levy, C. 1980. *Emancipation, sugar and federalism: Barbados and the West Indies, 1833–1876.* Gainesville: University Presses of Florida

Lewis, G. K. 1983. *Main currents in Caribbean thought: The historical evolution of Caribbean society in its ideological aspects, 1492–1900.* Baltimore; The Johns Hopkins University Press

    1987. *Grenada: The jewel despoiled.* Baltimore: The Johns Hopkins University Press

Liss, P. K. 1983. *Atlantic empires: The network of trade and revolution, 1713–1826.* Baltimore: The Johns Hopkins University Press

Lowenthal, A. F. 1972. *The Dominican intervention.* Cambridge: Harvard University Press

    1984. The insular Caribbean as a crucial test for US policy. In *The Caribbean challenge: US policy in a volatile region,* ed. H. M. Erisman. Boulder, Colorado: Westview Press, 183–97

Lowenthal, D. 1958. The West Indies chooses a capital. *Geographical Review* 48: 336–64

    1960a. Population contrasts in the Guianas. *Geographical Review* 50: 41–58

    1960b. The range and variation of Caribbean societies. *Annals New York Academy of Sciences,* 83: 786–95

    1961. Caribbean views of Caribbean land. *Canadian Geographer,* 5: 1–9

    1972. *West Indian societies.* London: Oxford University Press

    1987a. Foreword. In *Land and development in the Caribbean,* eds. J. Besson and J. Momsen. London: Macmillan

    1987b. Social features. In *Politics, security and development in small states,* eds. C. Clarke and T. Payne. London: Allen & Unwin

Lowenthal, D. and L. Comitas. 1962. Emigration and depopulation: Some neglected aspects of population geography. *The Geographical Review,* 52: 195–210

Lundahl, M. 1979. *Peasants and poverty: A study of Haiti.* New York: St. Martin Press

MacDonald, S. B. And A. L. Gastmann. 1984. Miterrand's headache: The French Antilles in the 1980s. *Caribbean Review*, 13, no. 2: 19–21

Maingot, A. P. 1984. Perceptions as realities: The United States, Venezuela, and Cuba in the Caribbean. In *Latin American nations in world politics*, eds. H. Munoz and J. S. Tulchin. Boulder, Colorado: Westview Press, 63–82

1989. Caribbean international relations. In *The modern Caribbean*, eds. F. W. Knight and C. A. Palmer. Chapel Hill: The University of North Carolina Press, 259–92

Maldonado-Denis, M. 1972. *Puerto Rico: A socio-historic interpretation.* Translated by Elena Vialo. New York: Random House

Mandle, J. R. 1985. *Big revolution, small country: The rise and fall of the Grenada revolution.* Lanham, Maryland: The North-South Publishing Company

Manley, M. 1975. *The politics of change: A Jamaican testament.* Washington: Howard University Press

Márquez, R. 1989. Nationalism, nation, and ideology: Trends in the emergence of a Caribbean literature. In *The modern Caribbean*, eds. F. W. Knight and C. A. Palmer. Chapel Hill: The University of North Carolina Press, 293–340

Marshall, B. A. 1976. Maronage in slave plantation societies: A case study of Dominica, 1785–1815. *Caribbean Quarterly*, 22: 26–32

Marshall, D. 1979. *The Haitian problem: Illegal migration to the Bahamas.* Mona, Jamaica: Institute of Social and Economic Research

Marshall, W. K. 1965. Metayage in the sugar industry of the British Windward islands, 1838–1865. *The Jamaican Historical Review*, 5: 28–55

1968. Peasant development in the West Indies since 1838. *Social and Economic Studies*, 17: 252–63

(ed.) 1977. *The Colthurst journal: Journal of a special magistrate in the islands of Barbados and St. Vincent, July 1835–September 1838.* New York: KTO Press

Matthews, H. L. 1975. *Revolution in Cuba: An essay in understanding.* New York: Charles Scribner's Sons

May, R. E. 1973. *The southern dream of a Caribbean empire.* Baton Rouge: Louisiana State University Press

McCullough, D. 1977. *The path between the seas: The creation of the Panama Canal, 1870–1914.* New York: Simon & Schuster

McFadden, R. D. 1989. Hurricane pounds isles in Caribbean. *New York Times*, September 19

McPherson, J. M. 1988. *Battle cry of freedom: The Civil War era.* New York: Oxford University Press

Meinig, D. W. 1986. *The shaping of America: A geographical perspective on 500 years of history.* Vol. 1, *Atlantic America, 1492–1800.* New Haven: Yale University Press

Mesa-Lago, C. 1981. *The economy of socialist Cuba.* Albuquerque: University of New Mexico Press

Mills, C. W. 1960. *Listen, Yankee: The revolution in Cuba.* New York: McGraw-Hill

Mills, F. L. 1974. The development of alternative farming systems and prospects for change in the structure of agriculture in St. Kitts, West Indies. Ph.D. dissertation, Clark University, Worcester, Massachusetts

Mims, S. L. 1912. *Colbert's West India Policy.* New Haven: Yale University Press

Mintz, S. W. 1953a. The culture history of a Puerto Rican sugar-cane plantation: 1876–1949. *The Hispanic American Historical Review,* 33: 224–51

1953b. The folk-urban continuum and the rural proletarian community. *American Journal of Sociology,* 59: 136–43

1964. Foreword to Guerra y Sánchez, R., *Sugar and society in the Caribbean: An economic history of Cuban agriculture.* New Haven: Yale University Press, xi–xliv

1974a. The Caribbean region, *Daedalus,* 103, no. 2: 45–71

1974b. *Caribbean transformations.* Chicago: Aldine

1977. The so-called world system: Local initiative and local response. *Dialectical Anthropology,* 2: 253–70

1985. *Sweetness and power: The place of sugar in modern history.* New York: Viking

Mintz, S. W. and R. Price. 1976. *An anthropological approach to the Afro-American past: A Caribbean perspective.* Philadelphia: Institute for the Study of Human Issues

Momsen, J. H. 1990. Changing patterns in Canada's trade with the Commonwealth Caribbean, 1966–1986. Paper presented at the meetings of the Society for Caribbean Studies, Hertfordshire, England, July, 25 pp. typescript

Moran, T. H. 1987. Cuban nickel development. In *Cuban Communism,* ed. I. L. Horowitz, 6th edn. New Brunswick, New Jersey: Transaction Books, 255–72

Moreno Fraginals, M. 1985. Plantations in the Caribbean: Cuba, Puerto Rico, and the Dominican Republic in the late nineteenth century. In *Between slavery and free labor: The Spanish-speaking Caribbean in the nineteenth century.* Baltimore: The Johns Hopkins University Press, 3–21

1976. *The Sugarmill: The socioeconomic complex of sugar in Cuba, 1760–1860.* Translated by C. Belfrage. New York: Monthly Review Press

Moya Pons, F. 1984. The Tainos of Hispaniola. *Caribbean Review,* 13, Fall: 20–23, 47

Munro, D. G. 1964. *Intervention and dollar diplomacy in the Caribbean, 1901–1921.* Princeton: Princeton University Press

Naipaul, V. S. 1984. An island betrayed. *Harper's,* March: 61–72

Nash, J. W. 1985. What hath intervention wrought: Reflections on the Dominican Republic. *Caribbean Review*, Fall: 6–11

Newson, L. A. 1976. *Aboriginal and Spanish colonial Trinidad: A study in culture contact.* London: Academic Press

Newton, V. 1984. *The silver men: West Indian labour migration to Panama, 1850–1914.* Mona, Jamaica: Institute of Social and Economic Research

Nietschmann, B. 1979. Ecological change, inflation, and migration in the far western Caribbean. *The Geographical Review*, 69: 1–24

Nistal-Moret, B. 1985. Problems in the social structure of slavery in Puerto Rico during the process of abolition, 1872. In *Between slavery and free labor: The Spanish-speaking Caribbean in the nineteenth century.* Baltimore: The Johns Hopkins University Press, 141–57

Olwig, K. F. 1985. *Cultural adaptation and resistance on St. John: Three centuries of Afro-Caribbean life.* Gainesville: University Presses of Florida

Ortiz, E. L. 1967. *The complete book of Caribbean cooking.* New York: M. Evans and Company

Ortiz, F. 1947. *Cuban counterpoint: Tobacco and sugar.* New York: Alfred A. Knopf

Ott, T. O. 1973. *The Haitian revolution, 1789–1804.* Knoxville: The University of Tennessee Press

Oxaal, I. 1982. *Black intellectuals and the dilemmas of race and class in Trinidad.* Cambridge, Massachusetts: Schenkman Publishing Company

Palmer, R. W. 1979. *Caribbean dependence on the United States economy.* New York: Praeger

Pares, R. 1936. *War and trade in the West Indies, 1739–1763.* London: Oxford University Press

1950. *A West-India fortune.* London: Longmans, Green & Co

Pastor, R. A. (ed.) 1985. *Migration and development in the Caribbean.* Boulder, Colorado: Westview Press

Patterson, O. 1978. Migration in Caribbean societies: Socioeconomic and symbolic resource. In *Human migration: Patterns and policies*, eds. W. H. McNeil and R. S. Adams. Bloomington: Indiana University press, 106–45

Patterson, S. 1969. *Immigration and race relations in Britain.* London: Oxford University Press

Pawson, M. and D. Buisseret. 1975. *Port Royal, Jamaica.* Oxford: Clarendon Press

Payne, A. J. 1980. *The politics of the Caribbean community, 1961–79: Regional integration amongst new states.* Manchester: Manchester University Press

1984. *The international crisis in the Caribbean.* London: Croom Helm

1985. Whither CARICOM? The performance and prospects of Caribbean integration in the 1980s. *International Journal*, 40, no. 2: 207–28

1989. Britain and the Caribbean. Paper presented at the meetings of the Latin American Studies Association in Miami, Florida, December. 33 pp. typescript

Payne, A. J., P. Sutton, and T. Thorndike. 1984. *Grenada: Revolution and invasion*. London: Croom Helm

Peach, C. 1968. *West Indian migration to Britain: A social geography*. London: Oxford University Press

Pérez, L. A. 1986. *Cuba under the Platt Amendment, 1902–1934*. Pittsburgh: University of Pittsburgh Press

    1989. *Lords of the mountain: Social banditry and peasant protest in Cuba, 1878–1918*. Pittsburgh: University of Pittsburgh Press

Perusek, G. 1984. Haitian emigration in the early twentieth century. *International Migration Review*, 18: 4–18

Petras. E. M. 1988. *Jamaican labor migration: White capital and black labor, 1850–1930*. Boulder, Colorado: Westview Press

Philpott, S. B. 1973. *West Indian migration: The Montserrat case*. London: Athlone Press

Pilkington, E. 1988. *Beyond the mother country: West Indians and the Notting Hill white riots*. London: I. B. Tauris

Pitt, D. E. 1989. Narcotics, not Nicaragua. *The New York Times Book Review*, January 29: 13

Plant, R. 1987. *Sugar and modern slavery. A tale of two countries*. London: Zed Books Ltd.

Plummer, B. G. 1988. *Haiti and the great powers, 1902–1915*. Baton Rouge: Louisiana State University Press

Polanyi-Levitt, K. 1985. The origins and implications of the Caribbean Basin Initiative: Mortgaging sovereignty? *International Journal*. 40, no. 2: 229–81

Pressley, S. A. 1988. Area Jamaicans struggle against drug-crime image. *Washington Post*, January 29: A1, A14

Prest, A. R. 1948. *War economies of primary producing countries*. Cambridge: Cambridge University Press

Preston, J. 1989. The trial that shook Cuba. *The New York Review of Books*, December 7: 24–31

Price, R. 1966. Caribbean fishing and fishermen: A historical sketch. *American Anthropologist*, 68: 1363–83

    (ed.) 1979. *Maroon societies: Rebel slave communities in the Americas*. 2nd edn., Baltimore: The Johns Hopkins University Press

    1983. *First-time. The historical vision of an Afro-American people*. Baltimore: The Johns Hopkins University Press

Proctor, J. H. 1962. British West Indian society and government in transition, 1920–60. *Social and Economic Studies*, 11, December: 273–304

Proudfoot, M. J. 1950. *Population movements in the Caribbean*. Port of Spain, Trinidad: Kent House

Quester, G. H. 1983. Trouble in the islands: Defending the micro-states. *International Security*, 8: 160–75

Ramsaran, R. 1989. *The Commonwealth Caribbean in the world economy.* London: Macmillan

Ratekin, M. 1954. The early sugar industry in Espanola, *The Hispanic American Historical Review*, 34: 1–19

Reid, I. D. A. 1939. *The Negro immigrant: His background, characteristics and social adjustment, 1899–1937.* New York: Columbia University Press

Richardson, B. C. 1972. Guyana's "green revolution": Social and ecological problems in an agricultural development programme. *Caribbean Quarterly*, 18: 14–23

1980. Freedom and migration in the leeward Caribbean, 1838–48. *Journal of Historical Geography*, 6: 391–408

1983. *Caribbean migrants: Environment and human survival on St. Kitts and Nevis.* Knoxville: The University of Tennessee Press

1984. Slavery to freedom in the British Caribbean: Ecological considerations. *Caribbean Geography*, 1: 164–75

1985. *Panama money in Barbados, 1900–1920.* Knoxville: The University of Tennessee Press

1987. Men, water, and mudflats in coastal Guyana. *Resource Management and Optimization* 5: 213–36

1989. Caribbean migrations, 1838–1985. In *The modern Caribbean*, eds. F. W. Knight and C. A. Palmer. Chapel Hill: The University of North Carolina Press, 203–28

Robbins, C. A. 1985. *The Cuban threat.* Philadelphia: Institute for the Study of Human Issues

Rodney, W. 1981. *A history of the Guyanese working people, 1881–1905.* Baltimore: The Johns Hopkins University Press

Rojas, E. 1984. Agricultural land in the eastern Caribbean: From resources for survival to resources for development. *Land Use Policy*, 1: 39–54

Rouse, I. 1986. *Migrations in prehistory.* New Haven: Yale University Press

Rubenstein, H. 1983. Remittances and rural underdevelopment in the English-speaking Caribbean. *Human Organization*, 42: 295–306

1987. *Coping with poverty: Adaptive strategies in a Caribbean village.* Boulder: Westview Press

Safa, H. I. 1974. *The urban poor of Puerto Rico: A study in development and inequality.* New York: Holt, Rinehart and Winston

Salas, L. P. 1987. Juvenile delinquency in postrevolutionary Cuba. In *Cuban Communism*, ed. I. L. Horowitz, 6th edn. New Brunswick, New Jersey: Transaction Books, 341–59

Sassen, S. 1988. *The mobility of labor and capital.* Cambridge: Cambridge University Press

Sauer, C. O. 1966. *The early Spanish main.* Berkeley: University of California Press

Schmidt, H. 1971. *The United States occupation of Haiti, 1915–1934.* New Brunswick: Rutgers University Press

Schwartz, R. 1989. *Lawless liberators: Political banditry and Cuban independence.* Durham, North Carolina: Duke University Press

Schwarzbeck, F. 1986. Guyane: A department like the others? In *Dual legacies in the contemporary Caribbean: Continuing aspects of British and French dominion,* ed. P. Sutton. London: Frank Cass

Scott, J. C. 1985. *Weapons of the weak: Everyday forms of peasant resistance.* New Haven: Yale University Press

Scott, R. J. 1985. *Slave emancipation in Cuba: The transition to free labor, 1860–1899.* Princeton: Princeton University Press

Searle, C. (ed.) 1984. *In nobody's backyard: Maurice Bishop's speeches, 1979–1983.* London: Zed Books

Sharp, R. H. 1942. In strategic Surinam. *Christian Science Monitor.* March 28: 4, 14

Sheridan, R. B. 1973. *Sugar and slavery: An economic history of the British West Indies, 1623–1775.* Baltimore: The Johns Hopkins University Press

1985. *Doctors and slaves: A medical and demographic history of slavery in the British West Indies, 1680–1834.* Cambridge: Cambridge University Press

Sigurdsson, H. 1982. In the volcano. *Natural History,* 91 (March): 60–66, 68

Silvestrini, B. G. 1989. Contemporary Puerto Rico: A society of contrasts. In *The modern Caribbean,* eds. F. W. Knight and C. A. Palmer. Chapel Hill: The University of North Carolina Press, 147–67

Simmons, D. A. 1985. Militarization of the Caribbean: Concerns for national and regional security. *International Journal,* 40, no. 2: 348–76

Singh, C. 1988. *Guyana: Politics in a plantation society.* New York: Praeger

Singham, A. W. 1968. *The hero and the crowd in a colonial polity.* New Haven: Yale University Press

Smith, R. T. 1956. *The Negro family in British Guiana: Family structure and social status in the villages.* London: Routledge & Kegan Paul

1962. *British Guiana.* London: Oxford University Press

Smith, W. S. 1987. *The closest of enemies: A personal and diplomatic account of US – Cuban relations since 1957.* New York: Norton

Solow, B. L. and S. L. Engerman (eds.) 1987. *British capitalism & Caribbean slavery.* Cambridge: Cambridge University Press

Stedman, J. G. 1988. *Narrative of a five years' expedition against the Bush Negroes of Surinam.* Eds. R. Price and S. Price. Baltimore: The Johns Hopkins University Press

Steele, P. 1988. *The Caribbean clothing industry: The US & Far East connections.* London: The Economist Intelligence Unit Ltd.

Stevens-Arroyo, A. M. 1984. A Taino tale: A mythological statement of social order. *Caribbean Review,* 13 (Fall): 24–26

Steward, J. H. *et al.* 1956. *The people of Puerto Rico.* Urbana: University of Illinois Press

Stone, C. 1988. Race and economic power in Jamaica: Toward the creation of a black bourgeoisie. *Caribbean Review,* 16 Spring: 10–14, 31–34

Sutton, P. (ed.) 1986. *Dual legacies in the contemporary Caribbean: Continuing aspects of British and French dominion.* London: Frank Cass

Sutton, P. 1987. Political aspects. In *Politics, security and development in small states,* eds. C. Clarke and T. Payne. London: Allen & Unwin

Taaffe, E. J., R. L. Morrill, and P. R. Gould. 1963. Transport expansion in underdeveloped countries: A comparative analysis. *Geographical Review,* 53: 503–29

Taylor, P. J. 1986. World-systems analysis. In *The dictionary of human geography.* 2nd edn. Ed. R. J. Johnston. Oxford: Blackwell, 527–29

1988. World-systems analysis and regional geography. *The Professional Geographer,* 40: 259–65

Thomas, C. Y. 1988. *The poor and the powerless: Economic policy and change in the Caribbean.* New York: Monthly Review Press

Thomson, R. 1987. *Green gold: Bananas and dependency in the eastern Caribbean.* London: Latin America Bureau

*Time.* 1959. Cuba: Vengeful visionary. January 26: 40–42, 47–49

Tinker, H. 1974. *A new system of slavery: The export of Indian labour overseas, 1830–1920.* London: Oxford University Press

Tomich, D. W. 1976. Prelude to emancipation: Sugar and slavery in Martinique, 1830–1848. Ph.D. dissertation (History), University of Wisconsin-Madison

Treaster, J. B. 1984. Jamaica, close US ally, does little to halt drugs. *The New York Times,* September 10: A1, A12

1985. Oil glut brings slump to Curacao and Aruba. *New York Times,* March 15: D1, D23

1988. 5 years later, Grenada is tranquil and thriving. *New York Times,* October 23

Trouillot, M. R. 1988. *Peasants and capital: Dominica in the world economy.* Baltimore: The Johns Hopkins University Press

Tuchman, B. W. 1988. *The first salute.* New York: Alfred A. Knopf

Wagley, C. 1957. Plantation America: A culture sphere. In *Caribbean studies: A Symposium,* ed. V. Rubin. Seattle: University of Washington Press, 3–13

Wallerstein, I. 1974. *The modern world-system I: Capitalist agriculture and the origins of the European world-economy in the sixteenth century.* New York: Academic Press

1980. *The modern world-system II: Mercantilism and the consolidation of the European world-economy, 1600–1750.* New York: Academic Press

1989. *The modern world-system III: The second era of great expansion of the capitalist world-economy, 1730–1840s.* New York: Academic Press

Watts, D. 1987. *The West Indies: Patterns of development, culture and environmental change since 1492.* Cambridge: Cambridge University Press

Weisskoff, R. 1985. *Factories and food stamps: The Puerto Rico model of development.* Baltimore: The Johns Hopkins University Press

West, R. C. And J. P. Augelli. 1976. *Middle America. Its lands and peoples.* 2nd edn. Englewood Cliffs, New Jersey: Prentice-Hall

Westlake, D. E. 1972. *Under an English heaven.* New York: Simon and Schuster

White, P. T. 1989. Coca. *National Geographic Magazine,* 175 January: 3–47

Wiarda, H. J. 1986–87. Misreading Latin America – again. *Foreign Policy,* 65: 135–53

Williams, E. 1944. *Capitalism and slavery.* Chapel Hill: The University of North Carolina Press

1970. *From Columbus to Castro: The history of the Caribbean, 1492–1969.* New York: Harper & Row

Wilson, P. J. 1973. *Crab antics: The Social anthropology of English-speaking Negro societies of the Caribbean.* New Haven: Yale University Press

Wolf, E. R. 1969. *Peasant wars of the twentieth century.* New York: Harper & Row

1982. *Europe and the people without history.* Berkeley: University of California Press

Wolf, E. R. and S. W. Mintz. 1957. Haciendas and plantations in Middle America and the Antilles. *Social and Economic Studies,* 6: 380–412

Wood, D. 1968. *Trinidad in transition: The years after slavery.* London: Oxford University Press

Worrell, D. 1987. *Small island economies: Structure and performance in the English-speaking Caribbean since 1970.* New York: Praeger

Young, A. 1958. *The approaches to local self-government in British Guiana.* London: Longmans, Green & Co

# Index